MEMOIRS
of the
American Mathematical Society
ISSN 0065-9266 (print)
ISSN 1947-6221 (online)

Number 1237

Fusion of Defects

Arthur Bartels
Christopher L. Douglas
André Henriques

March 2019 • Volume 258 • Number 1237 (first of 7 numbers)

AMS AMERICAN MATHEMATICAL SOCIETY

Library of Congress Cataloging-in-Publication Data

Cataloging-in-Publication Data has been applied for by the AMS. See http://www.loc.gov/publish/cip/.

DOI: https://doi.org/10.1090/memo/1237

Memoirs of the American Mathematical Society

This journal is devoted entirely to research in pure and applied mathematics.

Subscription information. Beginning with the January 2010 issue, *Memoirs* is accessible from www.ams.org/journals. The 2019 subscription begins with volume 257 and consists of six mailings, each containing one or more numbers. Subscription prices for 2019 are as follows: for paper delivery, US$1041 list, US$832.80 institutional member; for electronic delivery, US$916 list, US$732.80 institutional member. Upon request, subscribers to paper delivery of this journal are also entitled to receive electronic delivery. If ordering the paper version, add US$20 for delivery within the United States; US$80 for outside the United States. Subscription renewals are subject to late fees. See www.ams.org/help-faq for more journal subscription information. Each number may be ordered separately; *please specify number* when ordering an individual number.

Back number information. For back issues see www.ams.org/backvols.

Subscriptions and orders should be addressed to the American Mathematical Society, P. O. Box 845904, Boston, MA 02284-5904 USA. *All orders must be accompanied by payment.* Other correspondence should be addressed to 201 Charles Street, Providence, RI 02904-2213 USA.

Memoirs of the American Mathematical Society (ISSN 0065-9266 (print); 1947-6221 (online)) is published bimonthly (each volume consisting usually of more than one number) by the American Mathematical Society at 201 Charles Street, Providence, RI 02904-2213 USA. Periodicals postage paid at Providence, RI. Postmaster: Send address changes to Memoirs, American Mathematical Society, 201 Charles Street, Providence, RI 02904-2213 USA.

This publication is indexed in *Mathematical Reviews®*, *Zentralblatt MATH*, *Science Citation Index®*, *Science Citation Index™-Expanded*, *ISI Alerting Services℠*, *SciSearch®*, *Research Alert®*, *CompuMath Citation Index®*, *Current Contents®/Physical, Chemical & Earth Sciences*.

This publication is archived in *Portico* and *CLOCKSS*.

Printed in the United States of America.

∞ The paper used in this book is acid-free and falls within the guidelines established to ensure permanence and durability.

Visit the AMS home page at https://www.ams.org/

10 9 8 7 6 5 4 3 2 1 24 23 22 21 20 19

Contents

Abstract

Conformal nets provide a mathematical model for conformal field theory. We define a notion of defect between conformal nets, formalizing the idea of an interaction between two conformal field theories. We introduce an operation of fusion of defects, and prove that the fusion of two defects is again a defect, provided the fusion occurs over a conformal net of finite index. There is a notion of sector (or bimodule) between two defects, and operations of horizontal and vertical fusion of such sectors. Our most difficult technical result is that the horizontal fusion of the vacuum sectors of two defects is isomorphic to the vacuum sector of the fused defect. Equipped with this isomorphism, we construct the basic interchange isomorphism between the horizontal fusion of two vertical fusions and the vertical fusion of two horizontal fusions of sectors.

Received by the editor July 19, 2016.

Article electronically published on February 20, 2019.

DOI: https://doi.org/10.1090/memo/1237

2010 *Mathematics Subject Classification.* 81T05, 46L37, 46M05 (Primary), 81T40, 46L60, 81R10 (Secondary).

Key words and phrases. Conformal net, defect, sector, fusion, conformal field theory, soliton, vacuum, conformal embedding, Q-system, von Neumann algebra, Connes fusion.

This work first appeared as "Conformal nets III: Fusion of defects" at arXiv:1310:8263.

Acknowledgments

We are grateful for the guidance and support of Michael Hopkins, Michael Müger, Stephan Stolz, and Peter Teichner. AB was supported by the Sonderforschungsbereich 878, and CD was partially supported by a Miller Research Fellowship and by EPSRC grant EP/K015478/1. CD and AH benefited from the hospitality of MSRI (itself supported by NSF grant 0932078000) during its spring 2014 algebraic topology program.

Introduction

There are various different mathematical notions of field theories. For many of these there is also a notion of defects that formalizes interactions between different field theories. See for example [**FSV13**, **KS11**, **QRW07**, **SFR06**] and references therein. Depending on the context, sometimes the terminology 'surface operator' or 'domain wall' is used in place of 'defect'. Often field theories are described as functors from a bordism category whose objects are $(d-1)$-manifolds and morphisms are d-dimensional bordisms (usually with additional geometric structure) to a category of vector spaces. Defects allow the extension of such functors to a larger bordism category, where the manifolds may be equipped with codimension-1 submanifolds that split the manifolds into regions labeled by field theories. The codimension-1 submanifold itself is labeled by a defect between the field theories labeling the neighboring regions.

In this book we give a definition of defects for conformal nets. Conformal nets are often viewed as a particular model for conformal field theory.[1] Our main result is that under suitable finiteness assumptions there is a composition for defects that we call *fusion*. We also extend the notion of representations of conformal nets, also known as sectors, to the context of defects. Sectors between defects are a simultaneous generalization of the notion of representations of conformal nets, and of bimodules between von Neumann algebras. Ultimately, this will lead to a 3-category whose objects are conformal nets, whose 1-morphisms are defects, whose 2-morphisms are sectors, and whose 3-morphisms are intertwiners between sectors. The lengthy details of the construction of this 3-category are postponed to [**BDH16**], but the key ingredients of this 3-category will all be presented here. In [**BDH16**] we will use the language of internal bicategories developed in [**DH12**], but we expect that the results of the present book also provide all the essential ingredients to construct a 3-category of conformal nets, defects, sectors, and intertwiners in any other sufficiently weak model of 3-categories.

A. Conformal nets

Conformal nets grew out of algebraic quantum field theory and have been intensively studied; see for example [**BMT88**, **GF93**, **KL04**, **Was98**, **Was95**]. In this book we will use our (non-standard) coordinate-free definition of conformal nets [**BDH13**]. A conformal net in this sense is a functor

$$\mathcal{A}\colon \mathsf{INT} \to \mathsf{VN}$$

from the category of compact oriented intervals to the category of von Neumann algebras, subject to a number of axioms. The precise definition and properties of

[1] Note, however, that the precise relation of conformal nets to conformal field theory in the sense of Segal [**Seg04**] is not at present clear.

conformal nets are recalled in Appendix C. In contrast to the standard definition, in our coordinate-free definition there is no need to fix a vacuum Hilbert space at the outset—this feature will be useful in developing our definition of defects. Nevertheless, the vacuum Hilbert space can be reconstructed from the functor $\mathcal{A}$. The main ingredient for this reconstruction is Haagerup's standard form $L^2(A)$, a bimodule that is canonically associated to any von Neumann algebra A. This standard form and various facts about von Neumann algebras that are used throughout this book are reviewed in Appendix B.

B. Defects

To define defects we introduce the category $\mathsf{INT}_{\circ\bullet}$ of bicolored intervals. Its objects are intervals I that are equipped with a covering by two subintervals $I_\circ$ and $I_\bullet$. If I is not completely white ($I_\circ = I$, $I_\bullet = \emptyset$) or black ($I_\bullet = I$, $I_\circ = \emptyset$) then we require that the white and the black subintervals meet in exactly one point and we also require the choice of a local coordinate around this point. For conformal nets $\mathcal{A}$ and $\mathcal{B}$, a defect between them is a functor

$$D\colon \mathsf{INT}_{\circ\bullet} \to \mathsf{VN}$$

such that D coincides with $\mathcal{A}$ on white intervals and with $\mathcal{B}$ on black intervals and satisfies various axioms similar to those of conformal nets. Often we write ${}_{\mathcal{A}}D_{\mathcal{B}}$ to indicate that D is a defect from $\mathcal{A}$ to $\mathcal{B}$, also called an $\mathcal{A}$-$\mathcal{B}$-defect. A defect from the trivial net to itself is simply a von Neumann algebra (Proposition 1.22), so our notion of defect is a generalization of the notion of von Neumann algebra. The precise definition and some basic properties of defects are given in Chapter 1.

Certain defects have already appeared in disguise in the conformal nets literature, through the notion of 'solitons' [**BE98, Kaw02, LR95, LX04**]. For a conformal net $\mathcal{A}$ defined on subintervals I of the real line (half-infinite intervals allowed), an endomorphism of the C^*-algebra $\mathcal{A}_{\mathbb{R}} := \operatorname{colim} \mathcal{A}([a,b])$ is called a *soliton* if it is localized in a half-line, that is, if it acts as the identity on elements in the image of the complementary half-line. We bicolor any subinterval $I = [a,\infty] \cup [-\infty,b]$ of the projective line by $I_\circ = [a,\infty]$ and $I_\bullet = [-\infty,b]$. Given a soliton σ, we can consider the von Neumann algebra $D(I) := \sigma\mathcal{A}([a,\infty]) \vee \sigma\mathcal{A}([-\infty,b])$ generated by $\sigma\mathcal{A}([a,\infty])$ and $\sigma\mathcal{A}([-\infty,b])$ acting on the vacuum sector. We believe that, under certain conditions, this construction associates an $\mathcal{A}$-$\mathcal{A}$-defect to a soliton. The exact relationship between solitons and defects is, however, not yet clear.

C. Sectors

We will use the boundary of the square $S^1 := \partial[0,1]^2$ as our standard model for the circle. (The unit speed parametrization gives this circle a canonical smooth structure.) We equip the circle with the bicoloring defined by $S^1_\circ = S^1 \cap [0,\frac{1}{2}] \times [0,1]$ and $S^1_\bullet = S^1 \cap [\frac{1}{2},1] \times [0,1]$. Let D and E be $\mathcal{A}$-$\mathcal{B}$-defects. A D-E-sector is a Hilbert space equipped with compatible actions of the algebras $D(I)$, for subintervals $I \subset S^1$ with $(\frac{1}{2},0) \notin I$, and $E(I)$, for subintervals $I \subset S^1$ with $(\frac{1}{2},1) \notin I$. Pictorially we draw a D-E-sector as follows:

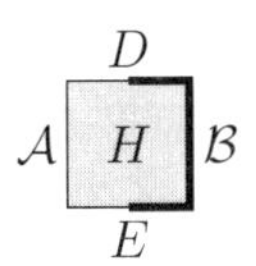

The thin line — should be thought of as white and stands for the conformal net $\mathcal{A}$ and the thick line ▬ should be thought of as black and stands for $\mathcal{B}$. Usually we simplify the picture further by dropping the letters. The precise definition and some basic properties of sectors are given in Chapter 2.

D. The vacuum sector of a defect

For any defect D we can evaluate D on the top half $S^1_\top :=$ ⌐¬ of the circle. Applying the L^2 functor we obtain the Hilbert space $H_0(S^1, D) := L^2(D(S^1_\top))$ and, as a consequence of the vacuum sector axiom in the definition of defects, this Hilbert space is a sector for D, called the vacuum sector for D. In our 3-category, the vacuum sector is the identity 2-morphism for the 1-morphism D. We often draw it as

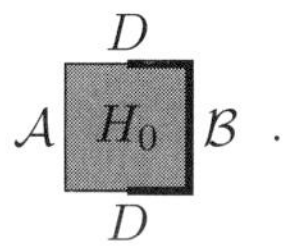

This darker shading is reserved for vacuum sectors.

E. Composition of defects

Let $D = {}_{\mathcal{A}}D_{\mathcal{B}}$ and $E = {}_{\mathcal{B}}E_{\mathcal{C}}$ be defects. Their composition or fusion $D \circledast_{\mathcal{B}} E$ is defined in Section 1.E. The definition is quite natural, but, surprisingly, it is not easy to see that $D \circledast_{\mathcal{B}} E$ satisfies all the axioms of defects.

We outline the definition of the fusion $D \circledast_{\mathcal{B}} E$. In our graphical notation, double lines ═ will now correspond to $\mathcal{A}$, thin lines — to $\mathcal{B}$, and thick lines ▬ to $\mathcal{C}$. Let us concentrate on the evaluation of $D \circledast_{\mathcal{B}} E$ on $S^1_\top$. Denote by $S^1_+ := \partial([1,2]\times[0,1])$ the translate of the standard circle and by $S^1_{+,\top}$ its top half. As with the vacuum sector $H_0(S^1, D) = L^2(D(S^1_\top))$ for D on S^1, we can form the vacuum sector $H_0(S^1_+, E) := L^2(E(S^1_{+,\top}))$ for E on S^1_+. Let $I = \{1\}\times[0,1]$ be the intersection of the two circles S^1 and S^1_+, equipped with the orientation inherited from S^1_+. The Hilbert space $H_0(S^1, D)$ is a right $\mathcal{B}(I)$-module, while $H_0(S^1_+, E)$ is a left $\mathcal{B}(I)$-module. Consequently, we can form their Connes fusion $H_0(S^1, D) \boxtimes_{\mathcal{B}(I)} H_0(S^1_+, E)$, drawn as

(0.1)

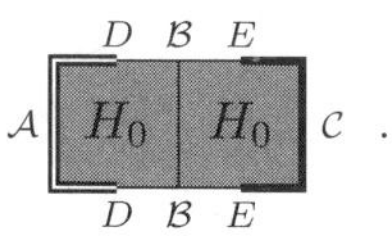

In this picture the middle vertical line | corresponds to I. By the axioms for defects the actions of $D(\ulcorner\!-)$ and of $\mathcal{B}(I)$ on $H_0(S^1, D)$ commute. Consequently, we obtain an action of $D(\ulcorner\!-)$ on the Connes fusion (0.1). Similarly, there is an action of $E(-\!\urcorner)$ on (0.1). Now $(D \circledast_{\mathcal{B}} E)(S^1_\top)$ is defined to be the von Neumann algebra generated by $D(\ulcorner\!-)$ and $E(-\!\urcorner)$ acting on the Hilbert space (0.1). A similar construction, using the local coordinate, is used to define the evaluation of $D \circledast_{\mathcal{B}} E$ on arbitrary bicolored intervals.

A main result of this book is that under the assumption that the intermediate conformal net $\mathcal{B}$ has finite index (see Appendix C.IV), the fusion of defects is in fact a defect:

THEOREM A (Existence of fusion of defects). *The fusion $D \circledast_{\mathcal{B}} E$ of two defects is again a defect.*

This is established in the text as Theorem 1.44.

F. Fusion of sectors and the interchange isomorphism

Let $\mathcal{A}$, $\mathcal{B}$, and $\mathcal{C}$ be conformal nets; let ${}_{\mathcal{A}}D_{\mathcal{B}}$, ${}_{\mathcal{B}}E_{\mathcal{C}}$, ${}_{\mathcal{A}}F_{\mathcal{B}}$, and ${}_{\mathcal{B}}G_{\mathcal{C}}$ be defects; and let $H =$  be a D–F-sector, and $K =$  an E–G-sector. The horizontal composition $H \boxtimes_{\mathcal{B}} K$ of these two sectors is a $(D \circledast_{\mathcal{B}} E)$–$(F \circledast_{\mathcal{B}} G)$-sector whose Hilbert space is the Connes fusion $H \boxtimes_{\mathcal{B}(I)} K$ and which is depicted as

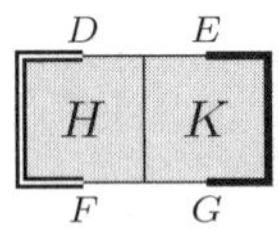

.

This composition operation is defined precisely in Section 2.B.

Let ${}_{\mathcal{A}}P_{\mathcal{B}}$ be another defect and let $L =$  be an F–P-sector. The vertical composition $H \boxtimes_F L$ of the sectors H and L is a D–P-sector whose Hilbert space is the Connes fusion $H \boxtimes_{F(S^1_\top)} L$. Here $S^1_\top$ is the top half of the circle bounding the sector L; this half circle is canonically identified, by vertical reflection, with the bottom half $S^1_\perp$ of the circle bounding the sector H, and that identification provides the action of $F(S^1_\top)$ on H. The vertical fusion operation occurs along half a circle, rather than a quarter circle as for horizontal fusion, and so is not conveniently depicted by juxtaposing squares; instead we denote this vertical composition by

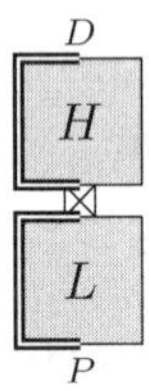

.

This composition operation is defined precisely in Section 2.C. Because vertical composition is Connes fusion along the algebra associated to half a circle, and the vacuum sector is defined as an L^2 space for the algebra associated to half a circle, the vacuum sector serves as an identity for vertical composition; that is, $H_0(S^1, D) \boxtimes_D H \cong H$ and $H \boxtimes_F H_0(S^1, F) \cong H$ as sectors.

Let ${}_{\mathcal{B}}Q_{\mathcal{C}}$ be yet another defect and let $M =$  be a G–Q-sector. The sectors K and M can be composed into an E–Q-sector $K \boxtimes_G M$. The horizontal composition and the vertical composition of sectors ought to be compatible in the sense that there is a canonical isomorphism

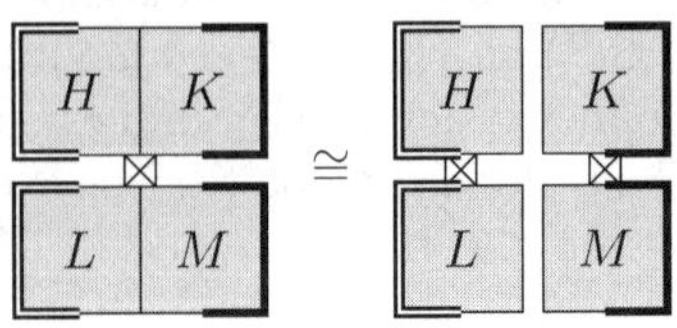

;

here, in the left picture the horizontal compositions occur first, followed by the vertical composition, whereas in the right picture, the vertical compositions occur first, followed by the horizontal composition. This fundamental "interchange isomorphism" is constructed in Section 6.D.

The construction leverages the special case of the isomorphism where all four sectors are vacuum sectors. That isomorphism is defined as the following composite,

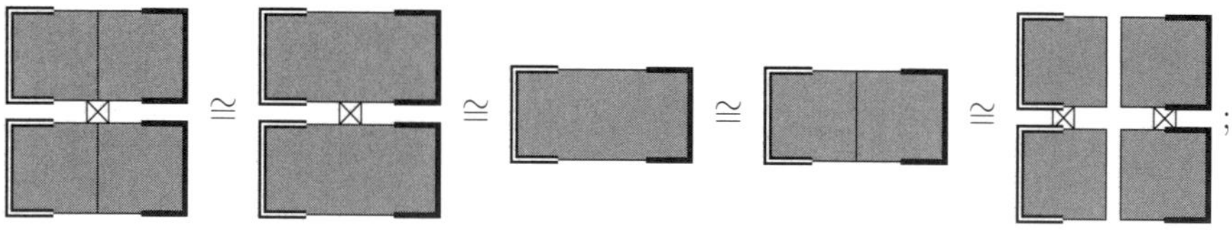

here the middle rectangle

(0.2)

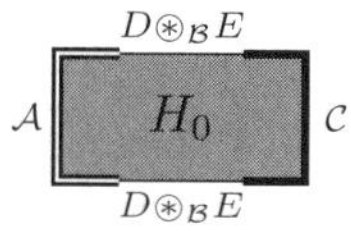

denotes the vacuum sector $H_0(S^1, D \circledast_{\mathcal{B}} E)$ of the composite defect $D \circledast_{\mathcal{B}} E$. The second and fourth isomorphisms in this composite are simply applications of the aforementioned property that the vacuum is an identify for vertical composition. The first and third isomorphisms are instances of the much more difficult fact that the horizontal composition of two vacuum sectors on defects is itself isomorphic to the vacuum sector of the composite defect. Because the horizontal composition is a Connes fusion, and the vacuum sector is a vertical unit, we call this isomorphism the "one times one isomorphism".

G. The $1 \boxtimes 1$-isomorphism

This isomorphism provides a canonical identification of the Hilbert space (0.1), used to define the defect $D \circledast_{\mathcal{B}} E$, with the vacuum sector for $D \circledast_{\mathcal{B}} E$. By definition the vacuum sector is $H_0(S^1, D \circledast_{\mathcal{B}} E) := L^2(D \circledast_{\mathcal{B}} E(S^1_\top))$.

By construction the algebra $D \circledast_{\mathcal{B}} E(S^1_\top)$ contains $D(\ulcorner)$ and $E(\urcorner)$ as two commuting subalgebras and is generated by those subalgebras. We can think of the algebra $D \circledast_{\mathcal{B}} E(S^1_\top)$ as associated to the tricolored interval $\ulcorner\!\!-\!\!\urcorner$ which is the upper half of the circle $\partial([0,2]\times[0,1])$; it is therefore natural, as above, to draw the vacuum sector for $D \circledast_{\mathcal{B}} E$ as $\square$.

Another main result of this book is the existence of an isomorphism between the fusion (0.1) and the vacuum (0.2). We presume as before that the net $\mathcal{B}$ has finite index.

THEOREM B (The $1 \boxtimes 1$-isomorphism). *There is a canonical isomorphism between the vacuum sector $H_0(D \circledast_{\mathcal{B}} E)$ of the fused defect $D \circledast_{\mathcal{B}} E$ and the Connes fusion $H_0(D) \boxtimes_{\mathcal{B}(I)} H_0(E)$ of the two vacuum sectors of the defects.*

This result appears in the text as Theorem 6.2. The construction of the $1 \boxtimes 1$-isomorphism is quite involved and is carried out in Chapters 4, 5, and 6. Chapter 6 also contains a short summary, on page 66, collecting all the necessary ingredients in one place.

H. Construction of the $1 \boxtimes 1$-isomorphism

For any von Neumann algebra A the standard form $L^2(A)$ carries commuting left and right actions of A, i.e., $L^2(A)$ is an A–A-bimodule. In the case of the vacuum sector $H_0(S^1, D) = L^2(D(\ulcorner\!\urcorner))$ these two actions correspond to the left actions of $D(⌜⌝)$ and of $D(⌞⌟)$.[2] One difficulty in understanding the Connes fusion (0.1) comes from the fact that the algebra $\mathcal{B}(I)$, over which the Connes fusion is taken, intersects both $D(⌜⌝)$ and $D(⌞⌟)$. To simplify the situation we will consider a variation of (0.1) with a hole in the middle,

(0.3) 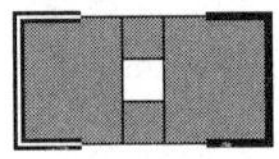 ;

we refer to this construction as keyhole fusion. This Hilbert space is built from vacuum sectors for D and E together with two (small) copies of the vacuum sector for $\mathcal{B}$. Its formal definition is given in Chapter 4; see in particular (4.8). The Connes fusion of $\mathcal{B}(I)$ is now replaced by four Connes fusion operations along smaller algebras. This allows us to identify, in Theorem 4.11, the Hilbert space (0.3) with the L^2-space of a certain von Neumann algebra that we represent by the graphical notation ⌜ ⍑ ⌝. It is generated by algebras $D(⌜)$, $\hat{\mathcal{B}}(⁼)$, and $E(⌝)$ acting on the Hilbert space ▮▮; here $\hat{\mathcal{B}}(⁼)$ is a certain enlargement of the algebra $\mathcal{B}(⁼)$ that we abbreviate graphically by ⁼. We defer to (4.9, 4.10) for the details of the definitions, and to 3.11 for an explanation of the notation $\hat{\mathcal{B}}$. Theorem 4.11 then reads

(0.4) 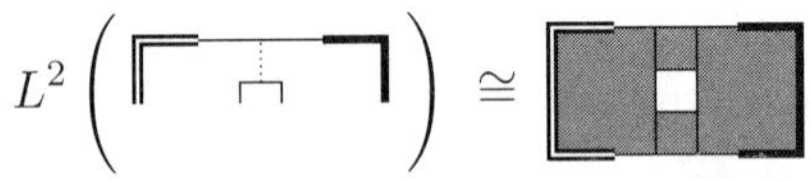

At this point we blur the distinction between intervals and algebras in our graphical notation and often draw only an interval to denote an algebra. For example we abbreviate $\mathcal{B}(⌐)$ as simply ⌐, and $D \circledast_{\mathcal{B}} E(S^1_\top)$ as ⌜ ⌝. We therefore write, for instance $D \circledast_{\mathcal{B}} E(S^1_\top) \otimes \mathcal{B}(⊓)$ as ⌜ ⊓ ⌝. As the notation indicates, this tensor product is a subalgebra of ⌜ ⍑ ⌝. (Note the additional dotted line in the middle.) If $\mathcal{B}$ has finite index, then we show in Corollary 4.16 that this inclusion ⌜ ⊓ ⌝ $\subseteq$ ⌜ ⍑ ⌝ is a finite homomorphism of von Neumann algebras. As the L^2-construction is functorial for such homomorphisms [**BDH11**], we can apply L^2 to it. Combining this with (0.4) we obtain a map

(0.5) 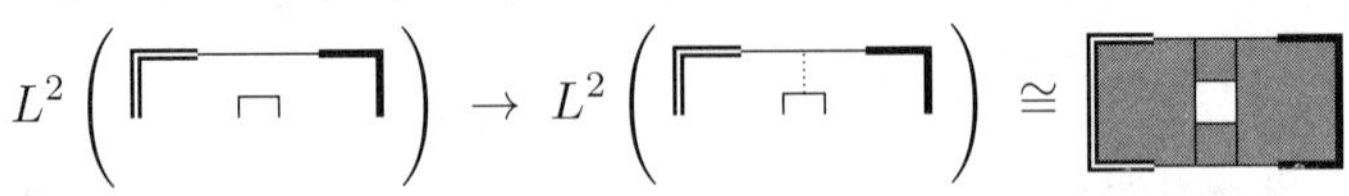

In the next step, we need to fill the keyhole in (0.3). Formally, this is done by applying Connes fusion with a further (small) vacuum sector for $\mathcal{B}$. We fancy this vacuum sector as the keystone and refer to the result as keystone fusion. On the domain of (0.5) the keystone cancels the algebra ⌐. On the target, we simply denote the result by filling the keyhole with the keystone, the (small) vacuum sector for $\mathcal{B}$. In this way we obtain, in Proposition 4.18, an isometric embedding

(0.6) 

The existences of this isometric embedding enables us to prove that $D \circledast_{\mathcal{B}} E$ is a defect. To produce the $1 \boxtimes 1$-isomorphism from (0.6), we construct, in Proposition

[2]The reflection along the horizontal axis $\mathbb{R}\times\{\frac{1}{2}\}$ provides an orientation reversing identification ⌜⌝ → ⌞⌟ and this accounts for the fact that the right action of $D(⌜⌝)$ on $L^2(D(⌜⌝))$ corresponds to a left action of $D(⌞⌟)$ on $H_0(S^1, D)$.

4.29, an isomorphism

(0.7) 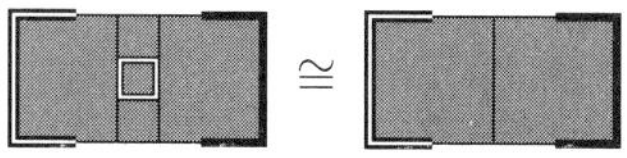

and then define the "$1 \boxtimes 1$-isomorphism" Ω as the composite of the two maps (0.6) and (0.7).

It remains to prove that the composite of (0.6) and (0.7) is indeed an isomorphism. The proof proceeds as follows: both the domain 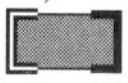 and the target 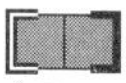 of Ω carry commuting actions of the algebras $(D \circledast_{\mathcal{B}} E)(S^1_\top) =$ 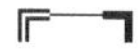 and $(D \circledast_{\mathcal{B}} E)(S^1_\bot) =$ . On 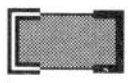 $= L^2($$)$ these two actions are clearly each other's commutants and so to prove that Ω is an isomorphism it suffices to show that the same holds for $H_0(D) \boxtimes_{\mathcal{B}(I)} H_0(E) =$ 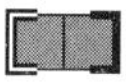. That these two actions are each other's commutants on this fusion of vacuum sectors, provided as before that the intermediate net has finite index, is a main technical results of this book:

THEOREM C (Haag duality for fusion of defects). *The algebras $(D \circledast_{\mathcal{B}} E)(S^1_\top)$ and $(D \circledast_{\mathcal{B}} E)(S^1_\bot)$, associated by the defect $D \circledast_{\mathcal{B}} E$ to the two halves of the circle, are each other's commutants in their action on the fusion $H_0(D) \boxtimes_{\mathcal{B}(I)} H_0(E)$ of the vacuum sectors of the defects.*

This is established in the text as Theorem 5.2; (see also Corollary 5.9). All of Chapter 5 is devoted to its proof.

REMARK. In constructing the 3-category of conformal nets, it is essential to know that the $1 \boxtimes 1$-isomorphism Ω satisfies certain axioms, such as associativity. In Proposition 4.32 we prove that the isomorphism is appropriately associative, but unfortunately this is done directly by tracing through the entire construction of Ω. Better would be to use a *characterization* of Ω (and thus of composites of multiple Ω maps) as the unique map satisfying certain properties. Haagerup's standard form (that is, the L^2-space of a von Neumann algebra) does admit such a characterization: it is determined up to unique unitary isomorphism by the module structure, the modular conjugation, and a self-dual cone. There is a natural choice of modular conjugation on 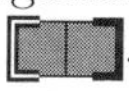. Thus, to characterize the isomorphism Ω, it suffices to specify a self-dual cone in that fusion of vacuum sectors. Unfortunately, we do not know how to construct such a self-dual cone from the self-dual cones of  and of .

CHAPTER 1

Defects

1.A. Bicolored intervals and circles

An interval is a smooth oriented 1-manifold diffeomorphic to $[0,1]$. We write $\mathrm{Diff}(I)$ for the group of diffeomorphisms of I and $\mathrm{Diff}_0(I)$ for the subgroup that fixes a neighborhood of ∂I. A *bicolored interval* is an interval I (always oriented) equipped with a cover by two closed, connected, possibly empty subsets $I_\circ, I_\bullet \subset I$ with disjoint interiors, along with a local coordinate (that is, an embedding $(-\varepsilon,\varepsilon) \hookrightarrow I$) at $I_\circ \cap I_\bullet$. We disallow the cases when $I_\circ$ or $I_\bullet$ consist of a single point. The local coordinate does *not* need to preserve the orientation, but is required to send $(-\varepsilon,0]$ into $I_\circ$ and $[0,\varepsilon)$ into $I_\bullet$. A bicolored interval necessarily falls in one of the following three classes:

(1.1) $I_\circ$, $I_\bullet$ are intervals and $I_\circ \cap I_\bullet$ is a point; the local coordinate is a smooth embedding $(-\varepsilon,\varepsilon) \hookrightarrow I$ that sends $(-\varepsilon,0]$ to $I_\circ$ and $[0,\varepsilon)$ to $I_\bullet$;
(1.2) $I_\circ = I$, $I_\bullet = \emptyset$, and there is no data of local coordinate;
(1.3) $I_\bullet = I$, $I_\circ = \emptyset$, and there is no data of local coordinate.

An embedding $f\colon J \hookrightarrow I$ from one bicolored interval to another is called color preserving if $f^{-1}(I_\circ) = J_\circ$ and $f^{-1}(I_\bullet) = J_\bullet$. The bicolored intervals form a category $\mathsf{INT}_{\circ\bullet}$, whose morphisms are the color preserving embeddings that respect the local coordinates (that is, such that the embedding intertwines the local coordinates on a sufficiently small neighborhood of 0). We let $\mathsf{INT}_\circ$ and $\mathsf{INT}_\bullet$ be the full subcategories on the objects of the form (1.2) and (1.3), respectively. Both of them are canonically isomorphic to INT, the category of uncolored intervals. Elements of $\mathsf{INT}_\circ$ and $\mathsf{INT}_\bullet$ are called *white* intervals and *black* intervals, respectively. Those of type (1.1) are called *genuinely bicolored* intervals. The full subcategory of genuinely bicolored intervals is denoted $\mathsf{INT}_{◑}$.

Similarly, a *bicolored circle* S is a circle (always oriented) equipped with a cover by two closed, connected, possibly empty subsets with disjoint interiors $S_\circ, S_\bullet \subset S$, along with local coordinates in the neighborhood of $S_\circ \cap S_\bullet$. We disallow the cases when $S_\circ$ or $S_\bullet$ consists of a single point. A bicolored circle necessarily falls in one of the following three categories:

(1.4) $S_\circ$ and $S_\bullet \subset S$ are intervals. The intersection $S_\circ \cap S_\bullet$ consist of two points and the two local coordinates are embeddings $(-\varepsilon,\varepsilon) \hookrightarrow S$ sending $(-\varepsilon,0]$ to $S_\circ$ and $[0,\varepsilon)$ to $S_\bullet$.
(1.5) $S_\circ = S$, $S_\bullet = \emptyset$, and there are no local coordinates;
(1.6) $S_\bullet = S$, $S_\circ = \emptyset$, and there are no local coordinates.

Bicolored circles of type (1.4) are called *genuinely bicolored.*

1.B. Definition of defects

Let VN be the category whose objects are von Neumann algebras with separable preduals, and whose morphisms are $\mathbb{C}$-linear homomorphisms, and $\mathbb{C}$-linear antihomomorphisms[3].

Recall our definition of conformal nets (see Appendix C). For the following definition of defect, we do not require that the conformal nets $\mathcal{A}$ and $\mathcal{B}$ are irreducible:

DEFINITION 1.7. Let $\mathcal{A}$ and $\mathcal{B}$ be two conformal nets. A defect from $\mathcal{A}$ to $\mathcal{B}$ is a functor

$$D\colon \mathsf{INT}_{\circ\bullet} \to \mathsf{VN}$$

that assigns to each bicolored interval I a von Neumann algebra $\mathcal{A}(I)$, and whose restrictions to $\mathsf{INT}_\circ$ and $\mathsf{INT}_\bullet$ are given by $\mathcal{A}$ and $\mathcal{B}$, respectively. It sends orientation-preserving embeddings to $\mathbb{C}$-linear homomorphisms, and orientation-reversing embeddings to $\mathbb{C}$-linear antihomomorphisms. The functor D is subject to the following axioms:

(i) *Isotony:* If I and J are genuinely bicolored intervals and $f : J \hookrightarrow I$ is an embedding, then $D(f) : D(J) \to D(I)$ is injective.
(ii) *Locality:* If $J \subset I$ and $K \subset I$ have disjoint interiors, then the images of $D(J)$ and $D(K)$ are commuting subalgebras of $D(I)$.
(iii) *Strong additivity:* If $I = J \cup K$, then the images of $D(J)$ and $D(K)$ topologically generate $D(I)$.
(iv) *Vacuum sector:* Let S be a genuinely bicolored circle, $I \subset S$ a genuinely bicolored interval, and $j\colon S \to S$ a color preserving orientation reversing involution that fixes ∂I. Equip $I' := j(I)$ with the orientation induced from S, and consider the following two maps of algebras:

$$\begin{aligned} \alpha \quad &: \quad D(I) \to \mathbf{B}(L^2D(I)) \\ \beta \quad &: \quad D(I') \xrightarrow{D(j)} D(I)^{op} \to \mathbf{B}(L^2D(I)) \end{aligned} \tag{1.8}$$

(Here α is the left action of $D(I)$ on $L^2D(I)$, and in β, the map $D(I)^{op} \to \mathbf{B}(L^2D(I))$ is the right action of $D(I)$ on $L^2D(I)$.) Let $J \in \mathsf{INT}_\circ \cup \mathsf{INT}_\bullet$ be a subinterval of I such that $J \cap \partial I$ consists of a single point, and equip $\bar{J} := j(J)$ with the orientation induced from S. We then require that the action

$$\alpha \otimes \beta : D(J) \otimes_{alg} D(\bar{J}) \longrightarrow \mathbf{B}\big(L^2D(I)\big) \tag{1.9}$$

of the algebraic tensor product extends to an action of $D(J \cup \bar{J})$.

The defect D is said to be *irreducible* if for every genuinely bicolored interval I, the algebra $D(I)$ is a factor. We will write ${}_{\mathcal{A}}D_{\mathcal{B}}$ to indicate that D is a defect from $\mathcal{A}$ to $\mathcal{B}$.

Note that in the above definition of a defect D, for an embedding $I \to J$ of a white or black interval I into a genuinely bicolored interval J, the induced map of von Neumann algebras $D(I) \to D(J)$ is not required to be an injection.

The following properties are consequences of the listed axioms and the corresponding properties of conformal nets: inner covariance (Proposition 1.10), the

[3]An antihomomorphism is a map satisfying $f(1) = 1$ and $f(ab) = f(b)f(a)$.

split property (Proposition 1.11), Haag duality (Proposition 1.16), and continuity (Proposition 1.21).

Inner covariance and the split property. Recall that $\mathrm{Diff}_0(I)$ is the subgroup of diffeomorphisms of I that fix some neighborhood of ∂I.

PROPOSITION 1.10 (Inner covariance for defects). *Let I be a genuinely bicolored interval, and let $\varphi \in \mathrm{Diff}_0(I)$ be a diffeomorphism that preserves the bicoloring and the local coordinate. Then $D(\varphi)$ is an inner automorphism of $D(I)$.*

PROOF. Write $\varphi = \varphi_\circ \circ \varphi_\bullet$ with $\mathrm{supp}(\varphi_\circ) \subset I_\circ$ and $\mathrm{supp}(\varphi_\bullet) \subset I_\bullet$. Let $\{J, K, L\}$ be a cover of I such that J is a white interval, K is a genuinely bicolored interval, L is a black interval, $\mathrm{supp}(\varphi_\circ)$ is contained in the interior of J, $\mathrm{supp}(\varphi_\bullet)$ is contained in the interior of L, and φ acts as the identity on K. By inner covariance for the nets $\mathcal{A}$ and $\mathcal{B}$ (see Appendix C.I), there are unitaries $u \in \mathcal{A}(J)$ and $v \in \mathcal{B}(L)$ that implement $\varphi_\circ$ and $\varphi_\bullet$. Let w be their product in $D(I)$. Then $waw^* = D(\varphi)a$ holds for every $a \in D(I)$ that is in the image of $\mathcal{A}(J)$, of $D(K)$, or of $\mathcal{B}(L)$. By strong additivity, it therefore holds for every element of $D(I)$. □

PROPOSITION 1.11 (Split property for defects). *If $J \subset I$ and $K \subset I$ are disjoint, then the map $D(J) \otimes_{alg} D(K) \to D(I)$ extends to the spatial tensor product $D(J) \,\bar{\otimes}\, D(K)$.*

PROOF. We assume without loss of generality that the interval J is entirely white and that it does not meet the boundary of I (otherwise, replace I by a slightly larger interval). Let $J^+ \subset I$ be a white interval that contains J in its interior and that does not intersect K. Finally, let $\iota : \mathcal{A}(J^+) \to D(I)$ be the map induced by the inclusion $J^+ \hookrightarrow I$. By the split property and Haag duality for conformal nets, the inclusion $\iota\,\mathcal{A}(J) \subseteq \iota\,\mathcal{A}(J^+)$ is split in the sense of Definition B.25. As $\mathcal{A}(J^+)$ commutes with $D(K)$, the inclusion $\iota\,\mathcal{A}(J) \to D(K)'$ is then also split, where the commutant is taken in any faithful representation of $D(I)$. Thus,

$$D(J) \otimes_{alg} D(K) = (\iota\,\mathcal{A}(J) \oplus \ker \iota) \otimes_{alg} D(K) \to D(I)$$

extends to the spatial tensor product $D(J) \,\bar{\otimes}\, D(K)$. □

Vacuum properties. Let S be a genuinely bicolored circle, along with an orientation reversing diffeomorphism $j \colon S \to S$, compatible with the bicoloring and with the local coordinates. Let $I \subset S$ be an interval whose boundary is fixed by j and let $I' := j(I)$. The Hilbert space $H_0 := L^2(D(I))$ is called the *vacuum sector* of D associated to S, I, and j. It is endowed with actions of $D(J)$ for every bicolored intervals $J \subset S$, as follows. (Recall that bicolored intervals contain at most one color-change point.) The maps (1.8) provide natural actions of $D(J)$ on H_0 for all subintervals $J \subset I$ and $J \subset I'$. By the vacuum sector axiom for defects, these extend to the algebras $D(J)$ associated to white and to black subintervals of S. To define the action $\rho_J : D(J) \to \mathbf{B}(H_0)$ of an arbitrary genuinely bicolored interval $J \subset S$, pick a white interval $K_1 \subset S$, a black interval $K_2 \subset S$, and diffeomorphisms $\varphi_i \in \mathrm{Diff}_0(K_i)$ such that $\varphi_1\varphi_2(J)$ does not cross ∂I. If $u_1 \in \mathcal{A}(K_1)$ and $u_2 \in \mathcal{B}(K_2)$ are unitaries implementing φ_1 and φ_2, then the action on H_0 of an element $a \in D(J)$ is defined by

$$\rho_J(a) := u_2^* u_1^* \,\rho_{\varphi_1\varphi_2(J)}\big(D(\varphi_1\varphi_2)(a)\big)\, u_1 u_2. \tag{1.12}$$

This action is compatible with the actions associated to other intervals, and is independent of the choices of φ_1, φ_2 and u_1, u_2 (see Lemma 2.5 for a similar construction in a more general context).

The following result, constructing isomorphisms between different vacuum sectors, is analogous to [**BDH13**, Cor. 1.16]:

LEMMA 1.13. *Let S be a genuinely bicolored circle. Let I_1 and I_2 be genuinely bicolored subintervals and let j_1 and j_2 be involutions fixing ∂I_1 and ∂I_2. Then the corresponding vacuum sectors $L^2D(I_1)$ and $L^2D(I_2)$ are non-canonically isomorphic as representations of the algebras $D(J)$ for $J \subset S$.*

PROOF. If I_1 and I_2 contain the same color-change point, then let $\varphi \in \mathrm{Diff}(S)$ be a diffeomorphism that sends I_1 to I_2, that intertwines j_1 and j_2, and that can be written as $\varphi = \varphi_\circ \circ \varphi_\bullet$ where $\varphi_\circ$ acts on the white part only and $\varphi_\bullet$ acts on the black part only. Let K be a white interval that contains $\mathrm{supp}(\varphi_\circ)$ in its interior and let L be a black interval that contains $\mathrm{supp}(\varphi_\bullet)$ in its interior. Finally, let $u \in \mathcal{A}(K)$ and $v \in \mathcal{B}(L)$ be unitaries implementing $\varphi_\circ$ and $\varphi_\bullet$. Then

$$L^2(D(I_1)) \xrightarrow{L^2(D(\varphi))} L^2(D(I_2)) \xrightarrow{u^*v^*} L^2(D(I_2))$$

is the desired isomorphism.

If I_1 and I_2 contain opposite color-change points, then we may assume without loss of generality that $j_1 = j_2$ and $I_2 = j_1(I_1)$. The isomorphism from $L^2(D(I_1))$ to $L^2(D(I_2))$ is then given by $L^2(D(j_1))$. □

NOTATION 1.14. Given a genuinely bicolored circle S and a defect ${}_{\mathcal{A}}D_{\mathcal{B}}$, we denote by $H_0(S, D)$ the vacuum sector associated to *some* interval $I \subset S$ and *some* involution j fixing ∂I. By the previous lemma, that representation of the algebras $D(J)$ (for $J \subset S$) is well defined up to non-canonical unitary isomorphism.

Our next result, concerning the gluing of vacuum sectors, is a straightforward generalization of [**BDH13**, Cor. 1.34] in the presence of defects (compare Appendix C.III). Let S_1 and S_2 be bicolored circles, let $I_i \subset S_i$ be bicolored intervals (whose boundaries do not touch the color change points), and let I_i' be the closure of $S_i \setminus I_i$. Assume that there exists an orientation reversing diffeomorphism $\varphi : I_2 \to I_1$ compatible with the bicolorings, and let $S_3 := I_1' \cup_{\partial I_2} I_2'$. Assume that $(S_3)_\circ$ and $(S_3)_\bullet$ are connected and non-empty. Then, up to exchanging S_1 and S_2, we are in one of the following three situations:

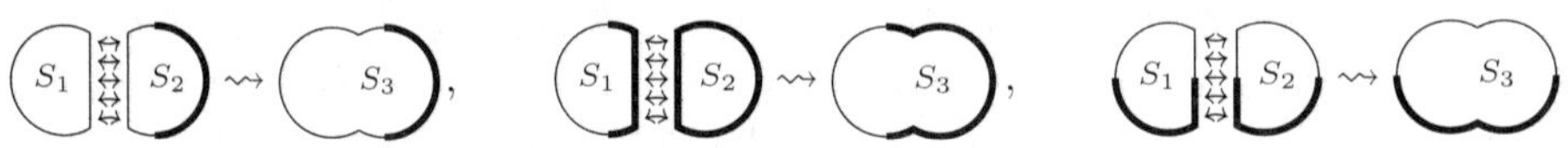

Equip $S_1 \cup_{I_2} S_2$ with a smooth structure that is compatible with the given smooth structures on S_1 and S_2 in the sense of [**BDH14**, Def. 1.4]. That is, provide smooth structures on S_1, S_2, and S_3 such that there exists an action of the symmetric group $\mathfrak{S}_3$ on $S_1 \cup_{I_2} S_2$ (with no compatibility with the bicoloring) that permutes the three circles and has $\pi|_{S_a}$ smooth for every $\pi \in \mathfrak{S}_3$ and $a \in \{1, 2, 3\}$.

When ${}_{\mathcal{A}}D_{\mathcal{B}}$ is a defect, it will be convenient to write $H_0(S, D) := H_0(S, \mathcal{A})$ if S is entirely white and $H_0(S, D) := H_0(S, \mathcal{B})$ if S is entirely black.

LEMMA 1.15. *Let S_1, S_2, S_3, and φ be as above, and let D be a defect. Use the map $D(\varphi)$ to equip $H_0(S_1, D)$ with the structure of a right $D(I_2)$-module. Then* *there exists a non-canonical isomorphism*

$$H_0(S_1, D) \boxtimes_{D(I_2)} H_0(S_2, D) \cong H_0(S_3, D),$$

compatible with the actions of $D(J)$ *for* $J \subset S_3$.

PROOF. Depending on the topology of the bicoloring, we can either identify $H_0(S_1, D)$ with $L^2(D(I_1))$ or identify $H_0(S_2, D)$ with $L^2(D(I_2))$. We assume without loss of generality that we are in the first case.

Let $j \in \mathrm{Diff}_-(S_1)$ be an involution that is compatible with the bicoloring and that fixes ∂I_1, and let $H_0(S_1, D) = L^2(D(I_1))$ be the vacuum sector associated to S_1, I_1, and j. We then have

$$L^2(D(I_1)) \underset{D(I_2)}{\boxtimes} H_0(S_2, D) \cong L^2(D(I_2)) \underset{D(I_2)}{\boxtimes} H_0(S_2, D) \cong H_0(S_2, D) \cong H_0(S_3, D),$$

where the first isomorphism uses $L^2(D(\varphi)) : L^2(D(I_2)) \to L^2(D(I_1)^{op}) = L^2(D(I_1))$ and the third one is induced by the map $(j \circ \varphi) \cup \mathrm{Id}_{I_2'} : S_2 \to S_3$. □

Haag duality. In certain cases, the geometric operation of complementation corresponds to the algebraic operation of relative commutant:

PROPOSITION 1.16 (Haag duality). *(1) Let* S *be a genuinely bicolored circle, let* $I \subset S$ *be a genuinely bicolored interval, and let* I' *be the closure of the complement of* I *in* S. *Then the algebras* $D(I)$ *and* $D(I')$ *are each other's commutants on* $H_0(S, D)$.

(2) Let ${}_{\mathcal{A}}D_{\mathcal{B}}$ *be a defect and let* $J \in \mathsf{INT}_{◑}$ *and* $K \in \mathsf{INT}_\circ \cup \mathsf{INT}_\bullet$ *be subintervals of* $I \in \mathsf{INT}_{◑}$. *Assume that* $J \cup K = I$ *and that* $J \cap K$ *is a point. Then* $D(J)$ *is the relative commutant of (the image of)* $D(K)$ *in* $D(I)$.

PROOF. (1) Let $j \in \mathrm{Diff}_-(S)$ be an involution that exchanges I and I' and that is compatible with the bicoloring and the local coordinates. By definition, we may take $H_0(S, D) = L^2(D(I))$ with the actions of $D(I)$ and $D(I')$ provided by (1.8). The result follows, as the left and right actions of $D(I)$ on $L^2(D(I))$ are each other's commutants.

(2) We assume without loss of generality that $K \in \mathsf{INT}_\circ$. Let $S := I \cup_{\partial I} (\bar{I})$ be a circle formed by gluing two copies of I along their boundary, such that there is a smooth involution j that exchanges them:

By strong additivity and the first part of the proposition, and considering actions on $H_0(S, D)$ we then have

$$D(K)' \cap D(I) = D(K)' \cap D(\bar{I})' = \big(D(K) \vee D(\bar{I})\big)' = D(K \cup \bar{I})' = D(J). \quad □$$

Canonical quantization. Let S be a bicolored circle and $I \subset S$ a genuinely bicolored interval. Let $j \in \mathrm{Diff}_-(S)$ be an involution that fixes ∂I and that is compatible with the bicoloring and the local coordinates. Also let $K \subset S$ be a white interval such that $j(K) = K$. We call a diffeomorphism $\varphi \in \mathrm{Diff}_0(K) \subset \mathrm{Diff}(S)$ *symmetric* if it commutes with j, and set

$$\mathrm{Diff}_0^{\mathrm{sym}}(K) := \{\varphi \in \mathrm{Diff}_0(K) \,|\, \varphi j = j\varphi\}.$$

Given a symmetric diffeomorphism φ, we also write $\varphi_0 \in \mathrm{Diff}(I)$ for $\varphi|_I$; to be precise, $\varphi_0 := \varphi|_{I\cap K} \cup \mathrm{id}_{I\setminus K}$.

For an irreducible defect ${}_{\mathcal{A}}D_{\mathcal{B}}$, we want to understand the automorphism $L^2D(\varphi_0)$ of $H_0(S,D) := L^2D(I)$, and its relation to the automorphism $L^2\mathcal{A}(\varphi_0)$ of $H_0(S,\mathcal{A}) := L^2\mathcal{A}(I)$, where in these expressions involving $\mathcal{A}$ the circle S has now been painted all white.

By [**BDH13**, Lem. 2.7] the unitary $u_\varphi := L^2\mathcal{A}(\varphi_0)$ on $L^2\mathcal{A}(I)$ implements φ, that is,

$$\mathcal{A}(\varphi)(a) = u_\varphi a {u_\varphi}^* \qquad \text{for all intervals } J \subseteq S \text{ and all } a \in \mathcal{A}(J). \tag{1.17}$$

Let K' be the closure of the complement of K in S. Since u_φ commutes with $\mathcal{A}(K')$, we have $u_\varphi \in \mathcal{A}(K)$ by Haag duality (Proposition C.4). We call $u_\varphi \in \mathcal{A}(K)$ the *canonical quantization of the symmetric diffeomorphism* φ.

The map $\mathrm{Diff}_0^{\mathrm{sym}}(K) \to \mathrm{Diff}_+(I)$ given by $\varphi \mapsto \varphi_0$ is continuous for the $\mathcal{C}^\infty$-topology. The map $\mathcal{A}\colon \mathrm{Diff}_+(I) \to \mathrm{Aut}(\mathcal{A}(I))$ is continuous because $\mathcal{A}$ is a continuous functor[4]. The map $\mathrm{Aut}(\mathcal{A}(I)) \to \mathrm{U}(L^2\mathcal{A}(I))$ given by $\psi \mapsto L^2(\psi)$ is continuous by [**Haa75**, Prop. 3.5]. Therefore, altogether, $\varphi \mapsto u_\varphi$ defines a continuous map from the group of symmetric diffeomorphisms of K to $\mathrm{U}(\mathcal{A}(K))$.

LEMMA 1.18. *Let S, I, K, φ, φ_0, and u_φ be as above, let ${}_{\mathcal{A}}D_{\mathcal{B}}$ be an irreducible defect, and let $H_0 := L^2D(I)$ be the vacuum sector of D associated to S, I, and j. Then, letting ρ_K be the action of $\mathcal{A}(K)$ on H_0 (given by the vacuum sector axiom), we have $L^2(D(\varphi_0)) = \rho_K(u_\varphi)$.*

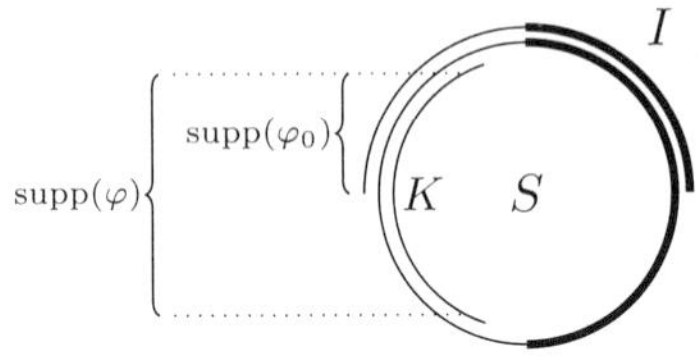

PROOF. We first show that the map

$$\begin{aligned} \mathrm{Diff}_0^{\mathrm{sym}}(K) &\to \mathrm{Aut}(D(I)) \\ \varphi &\mapsto D(\varphi_0) \end{aligned} \tag{1.19}$$

is continuous for the $\mathcal{C}^\infty$ topology on $\mathrm{Diff}_0^{\mathrm{sym}}(K)$ and the u-topology on $\mathrm{Aut}(D(I))$. Since $\mathrm{Ad}(u_\varphi) = \mathcal{A}(\varphi)$, the operator $\rho_K(u_\varphi)$ implements φ on H_0. In particular, $D(\varphi_0)$ is the restriction of $\mathrm{Ad}(\rho_K(u_\varphi))$ under the embedding $D(I) \hookrightarrow \mathbf{B}(H_0)$. The map

$$\mathrm{Diff}_0^{\mathrm{sym}}(K) \to \mathrm{U}(\mathcal{A}(K)) \to \mathrm{U}(H_0), \qquad \varphi \mapsto u_\varphi \mapsto \rho_K(u_\varphi)$$

is continuous and lands in the subgroup $\mathrm{N} := \{u \in \mathrm{U}(H_0) \,|\, uD(I)u^* = D(I)\}$. Since $D(\varphi_0) = \mathrm{Ad}(\rho_K(u_\varphi))$ and $\mathrm{Ad} : \mathrm{N} \to \mathrm{Aut}(D(I))$ is continuous [**BDH13**, A.18], the map (1.19) is therefore also continuous. Recalling [**Haa75**, Prop. 3.5] that $L^2 : \mathrm{Aut}(D(I)) \to \mathrm{U}(L^2D(I))$ is continuous, we have therefore shown that

$$\mathrm{Diff}_0^{\mathrm{sym}}(K) \to \mathbf{B}(H_0), \qquad \varphi \mapsto L^2(D(\varphi_0))$$

is a continuous homomorphism.

[4]This refers to Haagerup's u-topology on $\mathrm{Aut}(\mathcal{A}(I))$, see [**Haa75**, Def. 3.4] or [**BDH13**, Appendix].

Recall that $\rho_K(u_\varphi)$ implements φ. By the same argument as in [**BDH13**, Lem. 2.7], $L^2(D(\varphi_0))$ also implements φ. It follows that

$$L^2(D(\varphi_0)) = \lambda_\varphi \rho_K(u_\varphi)$$

for some scalar $\lambda_\varphi \in S^1$. Thus, we get a continuous map $\varphi \mapsto \lambda_\varphi$ from the group of symmetric diffeomorphisms of K into U(1). Our goal is to show that $\lambda_\varphi = 1$.

Let $J_\mathcal{A}$ and J_D be the modular conjugations on $L^2(\mathcal{A}(I))$ and $L^2(D(I))$, and let π_K be the natural action of $\mathcal{A}(K)$ on $L^2(\mathcal{A}(I))$. Since $\pi_K(u_\varphi) = L^2(\mathcal{A}(\varphi_0))$ commutes with $J_\mathcal{A}$, we have $J_\mathcal{A}\pi_K(u_\varphi)J_\mathcal{A} = \pi_K(u_\varphi)$. Combined with the fact that $J_\mathcal{A}$ implements j [**BDH13**, Lem. 2.5], this implies the equation

$$\mathcal{A}(j)(u_\varphi^*) = u_\varphi. \tag{1.20}$$

Applying ρ_K to (1.20), then, by a straightforward analog of [**BDH13**, Lem. 2.5], we learn that $J_D\,\rho_K(u_\varphi)\,J_D = \rho_K(u_\varphi)$. Since both $L^2(D(\varphi_0))$ and $\rho_K(u_\varphi)$ commute with J_D and since the latter is antilinear, the phase factor λ_φ must be real. It follows that $\lambda_\varphi \in \{\pm 1\}$.

To finish the argument, note that $\mathrm{Diff}_0^{\mathrm{sym}}(K)$ is connected and that $\{\pm 1\}$ is discrete. The map $\varphi \mapsto \lambda_\varphi$ being continuous, it must therefore be constant. □

Continuity. Given genuinely bicolored intervals I and J, and a neighborhood N of $I_\circ \cap I_\bullet$, let $\mathrm{Hom}^{(N)}(I,J)$ denote the set of embeddings $I \to J$ that preserve the local coordinate on the whole of N (this only makes sense if N is contained in the domain of definition of the local coordinate). We equip $\mathrm{Hom}_{\mathsf{INT}_{\circ\bullet}}(I,J) = \bigcup_N \mathrm{Hom}^{(N)}(I,J)$ with the colimit of the $\mathcal{C}^\infty$ topologies on $\mathrm{Hom}^{(N)}(I,J)$.

Given two von Neumann algebras A and B, the Haagerup u-topology on $\mathrm{Hom}_{\mathsf{VN}}(A,B)$ is the topology of pointwise convergence for the induced map on preduals [**BDH13**, Appendix].

Proposition 1.21 (Continuity for defects). *Let $D\colon \mathsf{INT}_{\circ\bullet} \to \mathsf{VN}$ be a defect. Then D is a continuous functor: for bicolored intervals I and J the map*

$$\mathrm{Hom}_{\mathsf{INT}_{\circ\bullet}}(I,J) \to \mathrm{Hom}_{\mathsf{VN}}(D(I),D(J))$$

is continuous with the above topology on $\mathrm{Hom}_{\mathsf{INT}_{\circ\bullet}}(I,J)$ *and with Haagerup's u-topology on*

$$\mathrm{Hom}_{\mathsf{VN}}(D(I),D(J)).$$

Proof. For every N as above, we need to show that the map $D : \mathrm{Hom}^{(N)}(I,J) \to \mathrm{Hom}_{\mathsf{VN}}(D(I),D(J))$ is continuous. We argue as in [**BDH13**, Lem 4.6]. Pick a bicolored interval K, and identify I and J with subintervals of K via some fixed embeddings into its interior. Given a generalized sequence $\varphi_i \in \mathrm{Hom}^{(N)}(I,J)$, $i \in \mathcal{I}$, with limit φ, and given a vector ξ in the predual of $D(J)$, we need to show that $D(\varphi_i)_*(\xi)$ converges to $D(\varphi)_*(\xi)$ in $D(I)_*$.

Let $\mathrm{Diff}_0^{(N)}(K)$ be the subgroup of diffeomorphisms of K that fix N and also fix a neighborhood of ∂K. Pick an extension $\hat\varphi \in \mathrm{Diff}_0^{(N)}(K)$ of φ, and let $\hat\varphi_{n,i} \in \mathrm{Diff}_0^{(N)}(K)$, $n \in \mathbb{N}$, be extensions of φ_i such that $\|\hat\varphi_{n,i} - \hat\varphi\|_{\mathcal{C}^n} < \|\varphi_i - \varphi\|_{\mathcal{C}^n}$, where $\|\ \|_{\mathcal{C}^n}$ is any norm that induces the $\mathcal{C}^n$ topology. Letting F be the filter on $\mathbb{N} \times \mathcal{I}$ generated by the sets $\{(n,i) \in \mathbb{N} \times \mathcal{I} \,|\, n \geq n_0, i \geq i_0(n)\}$ (see [**BDH13**, Lem 4.6]), then $F\text{-lim}\ \hat\varphi_{n,i} = \hat\varphi$ in the $\mathcal{C}^\infty$-topology.

Write $\mathrm{Diff}_0^{(N)}(K)$ as $\mathrm{Diff}_\circ \times \mathrm{Diff}_\bullet$, where $\mathrm{Diff}_\circ$ is the subgroup of $\mathrm{Diff}_0^{(N)}(K)$ consisting of diffeomorphisms whose support is contained in the white part, and

$\mathrm{Diff}_\bullet$ is the subgroup of diffeomorphisms whose support is contained in the black part. The continuity of (1.19) shows that the map $\mathrm{Diff}_0^{(N)}(K) \to \mathrm{Aut}(D(K)) : \psi \mapsto D(\psi)$ is continuous when restricted to either $\mathrm{Diff}_\circ$ or $\mathrm{Diff}_\bullet$. The composite

$$\mathrm{Diff}_0^{(N)}(K) = \mathrm{Diff}_\circ \times \mathrm{Diff}_\bullet \to \mathrm{Aut}(D(K)) \times \mathrm{Aut}(D(K)) \xrightarrow{mult.} \mathrm{Aut}(D(K))$$

is therefore also continuous. It follows that F-lim $D(\hat{\varphi}_{n,i}) = D(\hat{\varphi})$ in the u-topology on $\mathrm{Aut}(D(K))$. Given a lift $\hat{\xi} \in D(K)_*$ of ξ, the vectors $D(\hat{\varphi}_{n,i})_*(\hat{\xi})$ therefore converge to $D(\hat{\varphi})_*(\hat{\xi})$. Composing with the projection $\pi : D(K)_* \twoheadrightarrow D(I)_*$, it follows that $D(\varphi_i)_*(\xi) = \pi(D(\hat{\varphi}_{n,i})_*(\hat{\xi}))$ converges to $\pi(D(\hat{\varphi})_*(\hat{\xi})) = D(\varphi)_*(\xi)$. □

1.c. Examples of defects

Von Neumann algebras as defects, free boundaries, and defects coming from conformal embeddings. The trivial conformal net $\underline{\mathbb{C}}$ evaluates to $\mathbb{C}$ on every interval [**BDH13**, Eg. 1.3].

PROPOSITION 1.22. *There is a one-to-one correspondence (really an equivalence of categories) between $\underline{\mathbb{C}}$-$\underline{\mathbb{C}}$-defects and von Neumann algebras.*

PROOF. Given a von Neumann algebra A, the associated defect is

$$\underline{A}(I) := \begin{cases} \mathbb{C} & \text{if } I \in \mathsf{INT}_\circ \text{ or } I \in \mathsf{INT}_\bullet \\ A & \text{if } I \in \mathsf{INT}_{\circ\bullet} \text{ and the local coordinate is orientation preserving} \\ A^{op} & \text{if } I \in \mathsf{INT}_{\circ\bullet} \text{ and the local coordinate is orientation reversing} \end{cases}$$

where A^{op} denotes the opposite of A.

Conversely, let D be a $\underline{\mathbb{C}}$-$\underline{\mathbb{C}}$-defect. Given a bicolored interval I, the orientation reversing map $\mathrm{Id}_I : I \to -I$ identifies $D(-I)$ with $D(I)^{op}$, where $-I$ denotes I with opposite orientation. So we just need to show is that the restriction of D to the subcategory of genuinely bicolored intervals with orientation preserving maps (compatible with the local coordinates) is equivalent to a constant functor. By applying Proposition 1.16, we see that every embedding $J \to I$ between two such intervals induces an isomorphism $D(J) \to D(I)$.

To finish the proof, we need to check that $D(\phi) = \mathrm{Id}_{D(I)}$ for any $\phi : I \to I$. Pick a neighborhood $J \subset I$ of $I_\circ \cap I_\bullet$ on which ϕ is the identity. Then the two arrows $D(J) \to D(I)$ in the commutative diagram

$$\begin{array}{ccc} D(I) & \xrightarrow{\quad D(\phi) \quad} & D(I) \\ & \nwarrow^{\cong} \quad\quad {}^{\cong}\nearrow & \\ & D(J) & \end{array}$$

are equal to each other, showing that $D(\phi) = \mathrm{Id}$. □

PROPOSITION 1.23. *Let $\mathcal{A}$ be a conformal net. Then the functor $\overleftarrow{\mathcal{A}} : \mathsf{INT}_{\circ\bullet} \to \mathsf{VN}$ given by*

$$\overleftarrow{\mathcal{A}}(I) := \begin{cases} \mathcal{A}(I_\circ) & \text{if } I_\circ \neq \emptyset, \\ \mathbb{C} & \text{otherwise} \end{cases}$$

is an $\mathcal{A}$-$\underline{\mathbb{C}}$-defect.

PROOF. The axioms for defects follow immediately from the corresponding axioms for $\mathcal{A}$ (Appendix C.I). □

Conformal embeddings provide examples of defects. Recall that a morphism of conformal nets $\tau : \mathcal{A} \to \mathcal{B}$ is a called a *conformal embedding* [**BDH13**, §1.5] if

$$\mathrm{Ad}(u) = \mathcal{A}(\varphi) \quad \Rightarrow \quad \mathrm{Ad}(\tau_I(u)) = \mathcal{B}(\varphi)$$

for every diffeomorphism $\varphi \in \mathrm{Diff}_0(I)$ and unitary $u \in \mathcal{A}(I)$.

PROPOSITION 1.24. *Let $\mathcal{A}$ and $\mathcal{B}$ be conformal nets and let $\tau : \mathcal{A} \to \mathcal{B}$ be a conformal embedding. Then*

$$D_\tau(I) := \begin{cases} \mathcal{A}(I) & \text{for } I \in \mathsf{INT}_\circ \\ \mathcal{B}(I) & \text{for } I \in \mathsf{INT}_{\circ\!\!\!\bullet} \cup \mathsf{INT}_\bullet \end{cases} \tag{1.25}$$

is an $\mathcal{A}$-$\mathcal{B}$-defect.

PROOF. The axioms of isotony, locality, and vacuum sector follow directly from the corresponding axioms for $\mathcal{B}$. It remains to prove strong additivity. We need to show that

$$\tau_{[0,1]}(\mathcal{A}([0,1])) \vee \mathcal{B}([1,2]) = \mathcal{B}([0,2]).$$

For every point $x \in (0,1)$, pick a diffeomorphism $\varphi_x \in \mathrm{Diff}_0([0,2])$ sending 1 to x, and let $u_x \in \mathcal{A}([0,2])$ be a unitary implementing $\mathcal{A}(\varphi_x)$. Since τ is a conformal embedding, we then have $u_x b u_x^* = \mathcal{B}(\varphi)(b)$ for all $b \in \mathcal{B}([0,2])$. Moreover, since

$$u_x \in \mathcal{A}([0,2]) = \mathcal{A}([0,1]) \vee \mathcal{A}([1,2]) \subset \mathcal{A}([0,1]) \vee \mathcal{B}([1,2])$$

and since u_x conjugates $\mathcal{B}([1,2])$ to $\mathcal{B}([x,2])$, we have $\mathcal{B}([x,2]) \subset \mathcal{A}([0,1]) \vee \mathcal{B}([1,2])$. The argument being applicable to any $x \in (0,1)$, it follows from [**BDH13**, Lem. 1.4] that

$$\mathcal{B}([0,2]) = \bigvee_{x \in (0,1)} \mathcal{B}([x,2]) \subset \mathcal{A}([0,1]) \vee \mathcal{B}([1,2]). \qquad \square$$

Defects from Q-systems. Longo and Rehren [**Lon94, LR95**] showed that given a conformal net $\mathcal{A}$ and a unitary Frobenius algebra object in the category of $\mathcal{A}$-sectors, one can construct an extension $\mathcal{A} \subset \mathcal{B}$, where $\mathcal{B}$ is a possibly non-local conformal net.

Here, the unitary Frobenius algebra object is an $\mathcal{A}$-sector A along with unit and multiplication maps $\eta : H_0 \to A$, $\mu : A \boxtimes A \to A$ subject to the relations

$$\begin{aligned} &\mu(\eta \boxtimes \mathrm{id}_A) = \mu(1 \boxtimes \eta) = \mathrm{id}_A && \text{(unitality)} \\ &\mu(\mu \boxtimes \mathrm{id}_A) = \mu(\mathrm{id}_A \boxtimes \mu) && \text{(associativity)} \\ &\mu^*\mu = (\mathrm{id}_A \boxtimes \mu)(\mu^* \boxtimes \mathrm{id}_A) && \text{(Frobenius),} \end{aligned}$$

and the normalization $\mu\mu^* = \eta^*\eta \cdot \mathrm{id}_A$.[5] Here, H_0 is the vacuum sector of $\mathcal{A}$ (the identity sector on the identity defect; see Section 2.A) and $\boxtimes$ is the operation of vertical fusion (see Section 2.C).

If one encodes, as is usually done in the literature, the $\mathcal{A}$-sector A by a localized endomorphism $\theta : \mathcal{A}([0,1]) \to \mathcal{A}([0,1])$,[6] then a unitary Frobenius structure on A

[5]This last condition is only appropriate for simple Frobenius algebra objects. The Q-systems considered below will all correspond to simple Frobenius algebra objects.

[6]Here, 'localized' means that θ restricts to the identity map on $\mathcal{A}([0,\epsilon])$ and $\mathcal{A}([1-\epsilon,1])$. One recovers the sector A from the endomorphism θ by letting the underlying Hilbert space of A be H_0 and twisting the action on the top half of the circle (identified with $[0,1]$) by θ.

can be specified by a choice of two elements $w, x \in \mathcal{A}([0,1])$ (w is the unit and x^* is the multiplication) satisfying the relations [**BKLR14**, (4.1)]:

$$\begin{aligned} wa = \theta(a)w,\ x\theta(a) = \theta^2(a)x,\ \forall a \in \mathcal{A}([0,1]) &\quad \text{(source and target of } w \text{ and } x)\\ w^*x = \theta(w^*)x = 1 &\quad \text{(unitality)}\\ xx = \theta(x)x &\quad \text{(associativity)}\\ xx^* = \theta(x^*)x &\quad \text{(Frobenius)}\\ w^*w = x^*x = d \cdot 1 &\quad \text{(normalization)} \end{aligned}$$

for some scalar d. A triple (θ, w, x) subject to this set of equations is called a Q-system [**Lon94**].

We now use the Q-system (θ, w, x) to construct an $\mathcal{A}$-$\mathcal{A}$-defect D. Given a genuinely bicolored interval I, we use the local coordinate to construct a new interval $I^+ := I_\circ \cup [0,1] \cup I_\bullet$. As θ is localized, there is a unique extension θ^+ of θ to $\mathcal{A}(I^+)$. It is determined by requiring $\theta^+(a) = a$ for $a \in \mathcal{A}(K)$ with $K \subset I^+ \setminus (0,1)$ and $\theta^+(a) = \theta(a)$ for $a \in \mathcal{A}([0,1])$. We define $D(I)$ as the algebra generated by $\mathcal{A}(I^+)$ and one extra element v, subject to the relations [**BKLR14**, (4.3)]:

$$\begin{gathered} va = \theta(a)v \quad \forall a \in \mathcal{A}(I^+), \\ v^* = w^*x^*v, \quad vv = xv, \quad w^*v = 1. \end{gathered} \tag{1.26}$$

The Longo–Rehren non-local extension $\mathcal{A} \subset \mathcal{B}$ associated to the Q-system is defined such that $\mathcal{B}(I^+) = D(I)$ [**LR95**].

PROPOSITION 1.27. *Let $\mathcal{A}$ be a conformal net, and let (θ, w, x) be a Q-system in $\mathcal{A}([0,1])$. Then D, as defined above, is a $\mathcal{A}$-$\mathcal{A}$-defect.*

PROOF. Isotony for D follows from isotony for $\mathcal{A}$ because the map $\mathcal{A}(I^+) \to D(I) : a \mapsto av$ is a bijection. Indeed, $D(I)$ could alternatively be defined to be the set $\{av | a \in \mathcal{A}(I^+)\}$ with unit element w^*v, multiplication $av \cdot bv := a\theta(b)xv$, and star operation $(av)^* := w^*x^*\theta(a^*)v$.

To prove locality, consider the situation where $K \subset I$ are genuinely bicolored, and J has disjoint interior from K. We need to show that $\mathcal{A}(J)$ and $D(K) = \mathcal{A}(K^+) \vee \{v\}$ commute inside $D(I)$. By locality for $\mathcal{A}$, the algebras $\mathcal{A}(J)$ and $\mathcal{A}(K^+)$ commute inside $\mathcal{A}(I^+)$. The algebra $\mathcal{A}(J)$ also commutes with v, because $va = \theta(a)v$ and $\theta|_{\mathcal{A}(J)} = \mathrm{id}$.

To prove strong additivity, consider the situation of a genuinely bicolored interval I that is the union of a black or white interval J and a genuinely bicolored interval K. By definition the algebra $D(I)$ is generated by $\mathcal{A}(I^+)$ and v, and similarly $D(K)$ is generated by $\mathcal{A}(K^+)$ and v. By the strong additivity of $\mathcal{A}$, we have

$$D(J) \vee D(K) = \mathcal{A}(J) \vee \mathcal{A}(K^+) \vee \{v\} = \mathcal{A}(I^+) \vee \{v\} = D(I).$$

We now address the vacuum sector axiom. Consider the situation of a genuinely bicolored interval I and a white subinterval J that touches one of the boundary points of I (the other case is identical). By the vacuum sector axiom for $\mathcal{A}$, it is enough to construct a unitary map

$$u : L^2D(I) \to L^2\mathcal{A}(I^+)$$

that is equivariant with respect to the left and right actions of $\mathcal{A}(J)$. The formula $E : D(I) \to \mathcal{A}(I^+)$, $E(av) := d^{-1} \cdot aw$, $a \in \mathcal{A}(I^+)$ defines a conditional expectation, and the corresponding orthogonal projection $p : L^2D(I) \to L^2\mathcal{A}(I^+)$ satisfies

$pbp^* = E(b)$ for $b \in D(I)$ [**Kos86**, Lem 3.2]. We claim that

$$u(\xi) := \sqrt{d} \cdot p(v\xi)$$

is the desired unitary map. The map u is both left and right $\mathcal{A}(J)$ equivariant because

$$u(a\xi a') = \sqrt{d}\cdot p(va\xi a') = \sqrt{d}\cdot p(\theta(a)v\xi a') = \sqrt{d}\cdot p(av\xi a') = \sqrt{d}\cdot ap(v\xi)a' = au(\xi)a'$$

for $a, a' \in \mathcal{A}(J)$. To check that it is unitary, we compute

$$uu^*(\xi) = d \cdot p(vv^*(p^*\xi)) = d \cdot E(vv^*)\xi \quad \text{and}$$

$$d \cdot E(vv^*) = d \cdot E(vw^*x^*v) = d \cdot E(\theta(w^*x^*)vv) = d \cdot E(\theta(w^*x^*)xv) =$$
$$= \theta(w^*x^*)xw = \theta(w^*)\theta(x^*)xw = \theta(w^*)xx^*w = 1$$

and

$$u^*u(\xi) = d \cdot v^*p^*p(v\xi) = d \cdot v^*ev\xi \quad \text{and}$$
$$v^*ev = v^*v_1v_1^*v = 1/d$$

where $e = p^*p$ is the Jones projection, and v_1, as in [**LR95**, Sec 2.5], satisfies $v_1v_1^* = e$ and $v^*v_1 = 1/\sqrt{d}$ (this last formula holds by substituting $v/\sqrt{d}$ in place of w, and v_1 in place of v, in [**LR95**, (2.14)]; see also [**Lon94**, (4.3)]). □

REMARK 1.28. The considerations in [**BKLR14**] suggest a generalization of Proposition 1.27 to the following situation. Let $\mathcal{A}$ be a conformal net together with three Q-systems (θ^L, w^L, x^L), (θ^R, w^R, x^R), and (θ, w, x). If (θ^L, w^L, x^L) and (θ^R, w^R, x^R) are in addition assumed to be commutative, then (1.26) can be used to define extensions $\mathcal{A} \subset \mathcal{B}^L$ and $\mathcal{A} \subset \mathcal{B}^R$, as in [**LR95**]. If (θ, w, x) is such that the corresponding non-local extension $\mathcal{C}$ contains $\mathcal{B}^L$ and $\mathcal{B}^R$ and, moreover, $\mathcal{B}^L$ is *left-local* with respect to $\mathcal{C}$ and $\mathcal{B}^R$ is *right-local*[7] with respect to $\mathcal{C}$, then the construction D as in Proposition 1.27 should define a $\mathcal{B}^L$-$\mathcal{B}^R$-defect.

Direct sums and direct integrals of defects.

LEMMA 1.29. *Let D_1 and D_2 be $\mathcal{A}$-$\mathcal{B}$-defects. Then their direct sum $E := D_1 \oplus D_2$ is also an $\mathcal{A}$-$\mathcal{B}$-defect. Here, E is defined by*

$$E(I) = D_1(I) \oplus D_2(I) \ \text{ for } I \in \mathsf{INT}_{◐}$$
$$E(I) = \mathcal{A}(I) \ \text{ for } I \in \mathsf{INT}_{\circ} \qquad E(I) = \mathcal{B}(I) \ \text{ for } I \in \mathsf{INT}_{\bullet}.$$

PROOF. The only non-trivial axiom is strong additivity. Consider the situation where $I = K \cup J$, with J genuinely bicolored and K white. Letting $\Delta : \mathcal{A} \to \mathcal{A}^{\oplus 2}$ denote the diagonal map, we need to show that $E(I)$ is equal to the subalgebra

$$\Delta\mathcal{A}(K) \vee E(J) \subset E(I)$$

generated by the images of $\Delta\mathcal{A}(K)$ and $E(J)$. (Note that our notation is a little bit misleading, as the map $\Delta\mathcal{A}(K) \to E(I)$ might fail to be injective). Pick a white interval $L \subset J$ that touches K in a point. Since Δ is a conformal embedding, it follows from the previous proof that $\Delta\mathcal{A}(K) \vee \mathcal{A}^{\oplus 2}(L) = \mathcal{A}^{\oplus 2}(K \cup L)$.

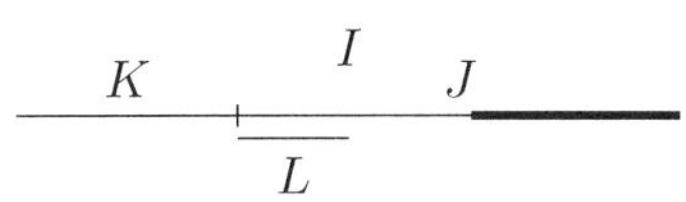

[7] This means that $\mathcal{B}^L(I)$ and $\mathcal{B}^R(I)$ should commute with $\mathcal{C}(J)$ whenever I is to the left, respectively to the right, of J.

Thus, we have the following equalities between subalgebras of $E(I)$:

$$\Delta\mathcal{A}(K)\vee E(J) = \Delta\mathcal{A}(K)\vee\mathcal{A}^{\oplus 2}(L)\vee E(J) = \mathcal{A}^{\oplus 2}(K\cup L)\vee E(J) = E(I).$$

□

REMARK 1.30. By the same argument as above, one can also show that a direct integral of $\mathcal{A}$-$\mathcal{B}$-defects is an $\mathcal{A}$-$\mathcal{B}$-defect.

Disintegrating defects. We show that defects between semisimple nets can be disintegrated. We warn the reader that our proof of the corresponding statement for conformal nets was incomplete:

CORRECTION 1.31. In [**BDH13**, §1.4, Eq 1.42], we claimed that every conformal net decomposes as a direct integral of irreducible ones. However, the group $\mathrm{Diff}(I)$ is not locally compact and it is not clear that its action on $\mathcal{A}(I)$ decomposes as a direct integral of actions on the irreducible components $\mathcal{A}(I)_x$ of $\mathcal{A}(I)$.[8] At present, we do not know how to fill this gap. This issue with [**BDH13**, Eq 1.42] does not affect any of the other results in [**BDH13**].

The above issue with disintegrating diffeomorphism actions does not arise here when disintegrating defects, because the relevant actions are inherited from the conformal nets.

Let D be a defect and let $f : J \to I$ be an embedding of genuinely bicolored intervals. Then one can show as follows that $D(f)$ induces an isomorphism between $Z(D(J))$ and $Z(D(I))$; compare the proof of [**BDH13**, Prop. 1.40]. We may as well assume that I and J share a boundary point. Let K be the closure of $I \setminus J$. The image of $Z(D(J))$ in $D(I)$ commutes with both $D(J)$ and $D(K)$, and so it is in $Z(D(I))$ by strong additivity. Conversely, $Z(D(I))$ commutes with $D(K)$, and is therefore contained in the image of $Z(D(J))$ by Haag duality (Proposition 1.16).

As in the case for conformal nets [**BDH13**, §1.4], we can then introduce an algebra $Z(D)$ that only depends on D, and that is canonically isomorphic to $Z(D(I))$ for every genuinely bicolored interval I. Disintegrating each $D(I)$ over that algebra, we can then write

$$D(I) = \int_{x\in X}^{\oplus} D_x(I) \quad \text{for every } I \in \mathsf{INT}_{\circ}$$

where X is any measure space with an isomorphism $L^\infty X \cong Z(D)$.

Recall that a conformal net is called semisimple if it is a finite direct sum of irreducible conformal nets (Appendix C.I). Similarly, we call a defect *semisimple* if it is a finite direct sum of irreducible defects.

LEMMA 1.32. *Any $\mathcal{A}$-$\mathcal{B}$-defect between semisimple conformal nets*[9] *is isomorphic to a direct integral of irreducible $\mathcal{A}$-$\mathcal{B}$-defects.*

PROOF. Fix a genuinely bicolored interval I. The algebra $D(I)$ disintegrates as explained above. We first need to show that for $K \subset I$ a white subinterval (respectively a black subinterval), the map $\mathcal{A}(K) \to D(I)$ (respectively $\mathcal{B}(K) \to D(I)$) similarly disintegrates. It suffices to see that $\mathcal{A}(K) \to D(I)$ induces maps $\mathcal{A}(K) \to D_x(I)$ for almost every x.

[8] We thank Sebastiano Carpi for pointing this out.

[9] Arbitrary direct sums of irreducible conformal nets would also work.

Note that it is in general not true that a map $N \to \int^{\oplus} M_x$ from a von Neumann algebra N (even a factor) into a direct integral induces maps $N \to M_x$ for almost every x. This is however true when N is a direct sum of type I factors. Indeed, letting $\mathcal{K} \subset N$ be the ideal of compact operators, we obtain maps $\mathcal{K} \to M_x$ by standard separability arguments. One then uses the fact that a C^*-algebra homomorphism from $\mathcal{K}$ into a von Neumann algebra extends uniquely to a von Neumann algebra homomorphism from N.

We can leverage this observation about direct sums of type I factors to construct the desired maps $\mathcal{A}(K) \to D_x(I)$. Consider a slightly larger interval I^+ that contains I, and let $K^+ \subset I^+$ be a white interval that contains K in its interior.

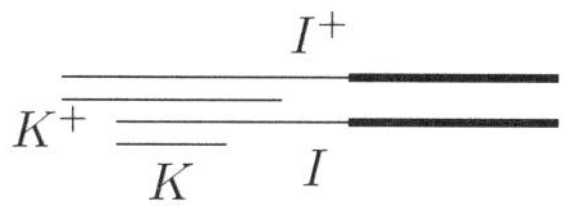

By the split property and the semisimplicity of $\mathcal{A}$, we can find an intermediate algebra $\mathcal{A}(K) \subset N \subset \mathcal{A}(K^+)$ that is a direct sum of type I factors. The map $N \to D(I^+) = \int^{\oplus} D_x(I^+)$ induces maps $\tilde{\iota}_x : N \to D_x(I^+)$ for almost every x; let ι_x denote the restriction of $\tilde{\iota}_x$ to $\mathcal{A}(K)$. The map $\int^{\oplus} \iota_x : \mathcal{A}(K) \to \int^{\oplus} D_x(I^+)$ is the composite of our original map $\mathcal{A}(K) \to D(I)$ with the inclusion $D(I) \hookrightarrow D(I^+)$. The image of $\int^{\oplus} \iota_x$ is contained in $\int^{\oplus} D_x(I)$. For almost every x the image of ι_x is therefore contained in $D_x(I)$, and we have our desired maps $\mathcal{A}(K) \to D_x(I)$.

Let $f \colon I \to I$ be an embedding of genuinely bicolored intervals. We need to know that $D(f)$ induces maps $D_x(I) \to D_x(I)$ for almost all x. As f fixes the local coordinate it suffices to consider the case where f is the identity outside of the interval K used above. We can then extend f to a diffeomorphism $f^+ \colon I^+ \to I^+$ that is the identity outside a small neighborhood of K. By inner covariance, $\mathcal{A}(f^+|_{K^+})$ is implemented by a unitary $u \in \mathcal{A}(K^+)$. The adjoint action $\mathrm{Ad}(u)$ induces the desired map $D_x(I) \to D_x(I)$. Since any genuinely bicolored interval J is isomorphic to I, we can transport the disintegration of the embedding $f \colon I \to I$ to a disintegration of any embedding of genuinely bicolored intervals.

The isotony, locality, and strong additivity axioms for D_x are immediate; the vacuum sector axiom requires a little bit more work, as follows. Let S, I, and J be as in the formulation of the vacuum sector axiom, and let us assume without loss of generality that J is white. We need to show that, for almost every x, the representation of $\mathcal{A}(J) \otimes_{alg} \mathcal{A}(\bar{J})$ on $H_0(S, D_x)$ extends to $\mathcal{A}(J \cup \bar{J})$. We know that the corresponding representation of $\mathcal{A}(J) \otimes_{alg} \mathcal{A}(\bar{J})$ on $H_0(S, D) = \int^{\oplus} H_0(S, D_x)$ does extend to $\mathcal{A}(J \cup \bar{J})$; we want to see that this extension disintegrates into actions of $\mathcal{A}(J \cup \bar{J})$ on $H_0(S, D_x)$. Certainly the action of $\mathcal{A}(J \cup \bar{J})$ on $H_0(S, D)$ commutes with that of $Z(D)$, but that is not enough to guarantee the action disintegrates into actions on the individual summands $H_0(S, D_x)$. Pick a white interval $K \subset S$ that contains $J \cup \bar{J}$ in its interior, and an intermediate algebra $\mathcal{A}(J \cup \bar{J}) \subset N \subset \mathcal{A}(K)$ that is a direct sum of type I factors. By the same argument as used earlier in this proof, the action of N on $\int^{\oplus} H_0(S, D_x)$ disintegrates into actions on $H_0(S, D_x)$, and therefore so does the action of $\mathcal{A}(J \cup \bar{J})$. □

Irreducible defects over semisimple nets. In Section 1.E, we will define the operation of fusion of defects, which is the composition of 1-morphisms in the 3-category of conformal nets. That operation does not preserve irreducibility (even if the conformal nets are irreducible) and so, unlike for conformal nets, it is not advisable to restrict attention to irreducible defects.

We call a defect D *faithful* if the homomorphisms $D(f)$ are injective for every embedding $f : I \to J$ of bicolored intervals.

LEMMA 1.33. *Let $\mathcal{A}$ and $\mathcal{B}$ be conformal nets, and let D be an $\mathcal{A}$-$\mathcal{B}$-defect that is irreducible and faithful. Then $\mathcal{A}$ and $\mathcal{B}$ are irreducible.*

PROOF. Let S be a genuinely bicolored circle, $I \subset S$ a white interval and I' the closure of its complement. Since D is irreducible, the vacuum sector $H_0(S, D)$ is acted on jointly irreducibly by the algebras $D(J)$, $J \subset S$.

Since D is faithful, $\mathcal{A}(I)$ acts faithfully on $H_0(S, D)$. A non-trivial central projection $p \in \mathcal{A}(I)$ would thus induce a non-trivial direct sum decomposition of $H_0(S, D)$, contradicting the fact that it is irreducible. Indeed, for a bicolored interval $J \subset S$, the projection p commutes with both $D(J \cap I)$ and $D(J \cap I')$. By strong additivity, p therefore commutes with $D(J)$. □

Here, as for conformal nets [**BDH13**, §3.1], we have used the split property to extend the functor D to disjoint unions of bicolored intervals by setting

$$D(I_1 \cup \ldots \cup I_n) := D(I_1) \bar{\otimes} \ldots \bar{\otimes} D(I_n). \tag{1.34}$$

COROLLARY 1.35. *Let $\mathcal{A} = \bigoplus \mathcal{A}_i$ and $\mathcal{B} = \bigoplus \mathcal{B}_j$ be semisimple conformal nets, where $\mathcal{A}_i$ and $\mathcal{B}_j$ are irreducible. Let D be an irreducible $\mathcal{A}$-$\mathcal{B}$-defect. Then there exist indices i and j such that D is induced from a faithful irreducible $\mathcal{A}_i$-$\mathcal{B}_j$-defect under the projections maps $\mathcal{A} \to \mathcal{A}_i$ and $\mathcal{B} \to \mathcal{B}_j$, respectively.* □

The above discussion shows that defects between semisimple conformal nets can be entirely understood in terms of defects between irreducible conformal nets. In the rest of this book, we will therefore mostly restrict attention to irreducible conformal nets.

1.D. The category CN_1 of defects

DEFINITION 1.36. Defects form a symmetric monoidal category CN_1. An object in that category is a triple $(\mathcal{A}, \mathcal{B}, D)$, where $\mathcal{A}$ and $\mathcal{B}$ are semisimple conformal nets, and D is a defect from $\mathcal{A}$ to $\mathcal{B}$. A morphism between the objects $(\mathcal{A}, \mathcal{B}, D)$ and $(\mathcal{A}', \mathcal{B}', D')$ is triple of natural transformations $\alpha\colon \mathcal{A} \to \mathcal{A}'$, $\beta\colon \mathcal{B} \to \mathcal{B}'$, $\delta\colon D \to D'$, with the property that $\delta|_{\mathsf{INT}_\circ} = \alpha$ and $\delta|_{\mathsf{INT}_\bullet} = \beta$. The symmetric monoidal structure on this category is given by objectwise spatial tensor product.

Recall that a map between von Neumann algebras with finite-dimensional centers is said to be finite if the associated bimodule ${}_AL^2B_B$ is dualizable (Appendix B.VI).

DEFINITION 1.37. A natural transformation $\tau : D \to E$ between semisimple defects ${}_{\mathcal{A}}D_{\mathcal{B}}$ and ${}_{\mathcal{C}}E_{\mathcal{D}}$ is called finite if $\tau_I : D(I) \to E(I)$ is a finite homomorphism for every $I \in \mathsf{INT}_{\circ\bullet}$.

REMARK 1.38. We believe that the condition of having finite-dimensional centers is not really needed to define the notion of finite homomorphism between von Neumann algebras [**BDH11**, Conj. 6.17]. If that is indeed the case, then we can extend the notion of finite natural transformations to non-semisimple defects.

Let us denote by CN_0 the symmetric monoidal category of semisimple conformal nets and their natural transformations (Appendix C.I), and by CN_0^f the symmetric monoidal category of semisimple conformal nets all of whose irreducible summands have finite index (Appendix C.IV), together with the finite natural transformations (Appendix C.I). Later on, we will denote by CN_1^f the symmetric monoidal category of semisimple defects (between semisimple conformal nets all of whose irreducible summands have finite index), together with finite natural transformations. The category CN_1 is equipped with two forgetful functors

$$\mathsf{source}\colon \mathsf{CN}_1 \to \mathsf{CN}_0 \qquad\qquad \mathsf{target}\colon \mathsf{CN}_1 \to \mathsf{CN}_0$$

given by $\mathsf{source}(\mathcal{A},\mathcal{B},D) := \mathcal{A}$ and $\mathsf{target}(\mathcal{A},\mathcal{B},D) := \mathcal{B}$, respectively, and a functor

$$\mathsf{identity}\colon \mathsf{CN}_0 \to \mathsf{CN}_1 \tag{1.39}$$

given by $\mathsf{identity}(\mathcal{A})(I) := \mathcal{A}(I)$, where we forget the bicoloring of I in order to evaluate $\mathcal{A}$. We sometimes abbreviate $\mathsf{identity}(\mathcal{A})$ by $1_{\mathcal{A}}$. Note that the above functors are all compatible with the symmetric monoidal structure.

REMARK 1.40. A conformal net $\mathcal{A}$ also has a *weak identity* given on genuinely bicolored intervals I by $I \mapsto \mathcal{A}(I_\circ \cup [0,1] \cup I_\bullet)$. That defect is not isomorphic to $1_{\mathcal{A}}$ in the category CN_1. It is nevertheless equivalent to $1_{\mathcal{A}}$ in the sense that there is an invertible sector between them; see Example 3.5.

1.E. Composition of defects

Given conformal nets $\mathcal{A}$, $\mathcal{B}$, $\mathcal{C}$, and defects ${}_{\mathcal{A}}D_{\mathcal{B}}$ and ${}_{\mathcal{B}}E_{\mathcal{C}}$, we will now define their fusion $D \circledast_{\mathcal{B}} E$, which is an $\mathcal{A}$-$\mathcal{C}$-defect if the conformal net $\mathcal{B}$ has finite index. If $\mathcal{B}$ does not have finite index, then $D \circledast_{\mathcal{B}} E$ might still be a defect, but we do not know how to prove this.

If I is in $\mathsf{INT}_\circ$ or $\mathsf{INT}_\bullet$, then $(D \circledast_{\mathcal{B}} E)(I)$ is given by $\mathcal{A}(I)$ or $\mathcal{C}(I)$, respectively. If I is genuinely bicolored, then we use the local coordinate to construct intervals

$$I^+ := I_\circ \cup [0,\tfrac{1}{2}], \quad I^{++} := I_\circ \cup [0,\tfrac{3}{2}], \quad {}^+I := [-\tfrac{1}{2},0] \cup I_\bullet, \quad {}^{++}I := [-\tfrac{3}{2},0] \cup I_\bullet,$$

bicolored by

$$\begin{aligned} I^+_\circ = I^{++}_\circ = I_\circ, \quad & I^+_\bullet = [0,\tfrac{1}{2}], \quad & I^{++}_\bullet = [0,\tfrac{3}{2}], \\ {}^+I_\bullet = {}^{++}I_\bullet = I_\bullet, \quad & {}^+I_\circ = [-\tfrac{1}{2},0], \quad & {}^{++}I_\circ = [-\tfrac{3}{2},0]. \end{aligned}$$

Let $J := [0,1]$ and consider the maps $J \to I^{++}_\bullet \hookrightarrow I^{++}$ and $J \to {}^{++}I_\circ \hookrightarrow {}^{++}I$ given by $x \mapsto \frac{3}{2} - x$ and $x \mapsto x - \frac{3}{2}$, respectively. These embeddings induce homomorphisms $D(I^{++}) \leftarrow \mathcal{B}(J)^{op}$ and $\mathcal{B}(J) \to E({}^{++}I)$ that we use to form the fusion of the von Neumann algebras $D(I^{++})$ and $E({}^{++}I)$ (Definition B.8). We define

$$(D \circledast_{\mathcal{B}} E)(I) := \begin{cases} \mathcal{A}(I) & \text{for } I \in \mathsf{INT}_\circ \\ D(I^{++}) \circledast_{\mathcal{B}(J)} E({}^{++}I) & \text{for } I \in \mathsf{INT}_{\circ\bullet} \\ \mathcal{C}(I) & \text{for } I \in \mathsf{INT}_\bullet \end{cases} \tag{1.41}$$

Pictorially, this is

(1.42) $(D \circledast_{\mathcal{B}} E)\Big($  $\Big) := D\Big($ 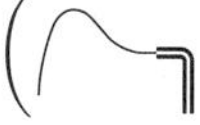 $\Big) \circledast_{\mathcal{B}(\Vert)} E\Big($  $\Big)$

If I is genuinely bicolored then, by Proposition 1.16, we have

$$\big(D \circledast_{\mathcal{B}} E\big)(I) = \big(D(I^{++}) \cap \mathcal{B}(J)^{op\prime}\big) \vee \big(E(^{++}I) \cap \mathcal{B}(J)'\big) = D(I^{+}) \vee E(^{+}I),$$

where the algebras act on $H \boxtimes_{\mathcal{B}(J)} K$ for some faithful $D(I^{++})$-module H and some faithful $E(^{++}I)$-module K. Therefore, we obtain the following equivalent definition of composition of defects:

Definition 1.43. The algebra $(D \circledast_{\mathcal{B}} E)(I)$ is the completion of the algebraic tensor product $D\big(I^{+}\big) \otimes_{alg} E\big(^{+}I\big)$ inside $\mathbf{B}(H \boxtimes_{\mathcal{B}(J)} K)$, where H is a faithful $D(I^{++})$-module, and K is a faithful $E(^{++}I)$-module.

We conjecture that that $D \circledast_{\mathcal{B}} E$ is always an $\mathcal{A}$-$\mathcal{C}$-defect. Our first main theorem says that this holds when $\mathcal{B}$ has finite index.

MAIN THEOREM 1.44. *Let $\mathcal{A}$, $\mathcal{B}$, and $\mathcal{C}$ be irreducible conformal nets, and let us assume that $\mathcal{B}$ has finite index. If D is a defect from $\mathcal{A}$ to $\mathcal{B}$, and E a defect from $\mathcal{B}$ to $\mathcal{C}$, then $D \circledast_{\mathcal{B}} E$ is a defect from $\mathcal{A}$ to $\mathcal{C}$.*

PROOF. We first prove isotony. Let $I_1 \subset I_2$ be genuinely bicolored intervals, let H be a faithful $D(I_2^{++})$-module and let K be a faithful $E(^{++}I_2)$-module. By the isotony property of D and E, the actions of $D(I_1^{++})$ on H and of $E(^{++}I_1)$ on K are faithful. Therefore, both $(D \circledast_{\mathcal{B}} E)(I_1)$ and $(D \circledast_{\mathcal{B}} E)(I_2)$ can be defined as subalgebras of $\mathbf{B}(H \boxtimes_{\mathcal{B}(J)} K)$. It is then clear that $(D \circledast_{\mathcal{B}} E)(I_1)$ is a subalgebra of $(D \circledast_{\mathcal{B}} E)(I_2)$.

We next show locality and strong additivity. Let $J \subset I$ and $K \subset I$ be bicolored intervals whose union is I and that intersect in a single point. We assume without loss of generality that K is white and that I and J are genuinely bicolored. In particular, we then have $^{+}I = {}^{+}J$. By the strong additivity of D, we have

$$\mathcal{A}(K) \vee (D \circledast_{\mathcal{B}} E)(J) = \mathcal{A}(K) \vee D(J^{+}) \vee E(^{+}J) = D(I^{+}) \vee E(^{+}I) = (D \circledast_{\mathcal{B}} E)(I),$$

which proves that $D \circledast_{\mathcal{B}} E$ is also strongly additive. Since D satisfies locality, the images of $\mathcal{A}(K)$ and $D(J^{+})$ commute in $D(I^{+})$. The algebra $D(I^{+})$ commutes with $E(^{+}I) = E(^{+}J)$ by the definition of $\circledast$. It follows that all three algebras $\mathcal{A}(K)$, $D(J^{+})$, and $E(^{+}J)$ commute with one another. The algebras $\mathcal{A}(K)$ and $(D \circledast_{\mathcal{B}} E)(J)$ therefore also commute, as required.

The vacuum sector axiom is much harder. Let us first assume that D and E are irreducible. Let $J \subset I$ be as in the formulation of the vacuum sector axiom (Definition 1.7), and let us assume without loss of generality that J is white. We need to show that the $\mathcal{A}(J) \otimes_{alg} \mathcal{A}(\bar{J})$-module structure on $L^2((D \circledast_{\mathcal{B}} E)(I))$ given by (1.8, 1.9) extends to an action of $\mathcal{A}(J \cup \bar{J})$. This will follow from the existence of an injective homomorphism from $L^2((D \circledast_{\mathcal{B}} E)(I))$ into some other $\mathcal{A}(J) \otimes_{alg} \mathcal{A}(\bar{J})$-module that is visibly an $\mathcal{A}(J \cup \bar{J})$-module. The desired homomorphism is (4.19) and will be constructed in Proposition 4.18. The fact that $\mathcal{A}(J \cup \bar{J})$ acts on the codomain of (4.19) is an immediate consequence of the vacuum sector axiom for D.

For general defects D and E, write them as direct integrals $D = \int^{\oplus} D_x$ and $E = \int^{\oplus} E_y$ of irreducible defects, and note that $D \circledast_{\mathcal{B}} E = \iint^{\oplus} D_x \circledast_{\mathcal{B}} E_y$ is a defect by Remark 1.30. □

In view of Corollary 1.35 and the fact that any defect between semisimple conformal nets can be disintegrated into irreducible defects (Lemma 1.32), the above theorem generalizes in a straightforward way to the situation where $\mathcal{A}$, $\mathcal{B}$, and $\mathcal{C}$ are not necessarily irreducible but merely semisimple: in this case, if all the irreducible summands of $\mathcal{B}$ have finite index, then the composition of an $\mathcal{A}$-$\mathcal{B}$-defect with a $\mathcal{B}$-$\mathcal{C}$-defect is an $\mathcal{A}$-$\mathcal{C}$-defect.

One might hope that composition of defects induces a functor

$$\mathsf{composition}\colon \mathsf{CN}_1 \times_{\mathsf{CN}_0} \mathsf{CN}_1 \to \mathsf{CN}_1. \tag{1.45}$$

However, some caution is needed. First, we used the finite index condition on $\mathcal{B}$ for our proof that $D \circledast_{\mathcal{B}} E$ is a defect. Second and more important, the operation of fusion of von Neumann algebras is only functorial with respect to isomorphisms of von Neumann algebras: given homomorphisms $A_1 \leftarrow C_1^{op}$, $C_1 \to B_1$ and $A_2 \leftarrow C_2^{op}$, $C_2 \to B_2$ it is *not* true that a triple of maps $a : A_1 \to A_2$, $b : B_1 \to B_2$, $c : C_1 \to C_2$ (subject to the obvious compatibility conditions) induces a map

$$a \circledast_c b : A_1 \circledast_{C_1} B_1 \to A_2 \circledast_{C_2} B_2. \tag{1.46}$$

Moreover, requiring that the maps a, b, and c be finite homomorphisms does not help to construct the map (1.46). However, unlike the fusion of von Neumann algebras, the composition of defects is functorial for more than just isomorphisms.

PROPOSITION 1.47. *Let $a : \mathcal{A}_1 \to \mathcal{A}_2$, $b : \mathcal{B}_1 \to \mathcal{B}_2$, and $c : \mathcal{C}_1 \to \mathcal{C}_2$ be natural transformations between irreducible conformal nets. Let ${}_{\mathcal{A}_1}D_1{}_{\mathcal{B}_1}$, ${}_{\mathcal{A}_2}D_2{}_{\mathcal{B}_2}$, ${}_{\mathcal{B}_1}E_1{}_{\mathcal{C}_1}$, and ${}_{\mathcal{B}_2}E_2{}_{\mathcal{C}_2}$ be defects, and let $d : D_1 \to D_2$, $e : E_1 \to E_2$ be natural transformations such that $d|_{\mathsf{INT}_\circ} = a$, $d|_{\mathsf{INT}_\bullet} = e|_{\mathsf{INT}_\circ} = b$, and $e|_{\mathsf{INT}_\bullet} = c$. If the natural transformation b is finite (Appendix C.I), then the above maps induce a natural transformation*

$$D_1 \circledast_{\mathcal{B}_1} E_1 \to D_2 \circledast_{\mathcal{B}_2} E_2.$$

Moreover, if $\mathcal{B}_i$ have finite index, D and E are semisimple, and d and e are finite, then the defects $D_i \circledast_{\mathcal{B}_i} E_i$ are semisimple and the above natural transformation is finite.

PROOF. Given a genuinely bicolored interval I, we need to construct a homomorphism $(D_1 \circledast_{\mathcal{B}_1} E_1)(I) \to (D_2 \circledast_{\mathcal{B}_2} E_2)(I)$. We assume without loss of generality that d and e are faithful (otherwise, their kernels are direct summands). Let H be a faithful $D_2(I^{++})$-module, and let K be a faithful $E_2({}^{++}I)$-module. By [**BDH11**, Thm. 6.23], the natural transformation b induces a bounded linear map $H \boxtimes_{\mathcal{B}_1(J)} K \to H \boxtimes_{\mathcal{B}_2(J)} K$, which is surjective by construction. That map is equivariant with respect to the homomorphism $D_1(I^+) \otimes_{alg} E_1({}^+I) \to D_2(I^+) \otimes_{alg} E_2({}^+I)$, and therefore induces a map from the completion of $D_1(I^+) \otimes_{alg} E_1({}^+I)$ in $\mathbf{B}(H \boxtimes_{\mathcal{B}_1(J)} K)$ to the completion of $D_2(I^+) \otimes_{alg} E_2({}^+I)$ in $\mathbf{B}(H \boxtimes_{\mathcal{B}_2(J)} K)$.

Let us now assume that the conformal nets $\mathcal{B}_i$ have finite index, the defects D and E are semisimple, and the natural transformations d and e are finite. The semisimplicity of $D_i \circledast_{\mathcal{B}_i} E_i$ is then the content of Theorem 3.6, and the homomorphism $(D_1 \circledast_{\mathcal{B}_1} E_1)(I) \to (D_2 \circledast_{\mathcal{B}_2} E_2)(I)$ is finite by [**BDH11**, Lem. 7.18]. □

We do not know whether the functor (1.45) exists as stated. However, instead of trying to compose over the full category CN_0 of semisimple conformal nets, we can restrict attention to the subcategory $\mathsf{CN}_0^f \subset \mathsf{CN}_0$ of semisimple conformal nets all of whose irreducible summands have finite index, together with their finite natural transformations. If we let $\mathsf{CN}_1 \times_{\mathsf{CN}_0^f} \mathsf{CN}_1$ be a shorthand notation for $\mathsf{CN}_1 \times_{\mathsf{CN}_0} \mathsf{CN}_0^f \times_{\mathsf{CN}_0} \mathsf{CN}_1$, then the composition functor

$$\text{(1.48)} \qquad \mathsf{composition}: \ \mathsf{CN}_1 \times_{\mathsf{CN}_0^f} \mathsf{CN}_1 \to \mathsf{CN}_1$$

exists by Theorem 1.44 and Proposition 1.47.

1.F. Associativity of composition

It will be convenient to work with the square model $S^1 := \partial[0,1]^2$ of the "standard circle" (see the beginning of Chapter 2) and to use the following notation.

NOTATION 1.49. Given real numbers $a < b$ and $c < d$ and $M = [a,b]\times[c,d]$, we let

$$\begin{aligned}
\partial^{\sqsubset} M &:= (\{a\}\times[c,d]) \cup ([a,b]\times\{c,d\}) & \partial^{\ulcorner} M &:= (\{a\}\times[c,d]) \cup ([a,b]\times\{d\})\\
\partial^{\sqsupset} M &:= (\{b\}\times[c,d]) \cup ([a,b]\times\{c,d\}) & \partial^{\llcorner} M &:= (\{a\}\times[c,d]) \cup ([a,b]\times\{c\})\\
\partial^{\sqcap} M &:= (\{a,b\}\times[c,d]) \cup ([a,b]\times\{d\}) & \partial^{\urcorner} M &:= (\{b\}\times[c,d]) \cup ([a,b]\times\{d\})\\
\partial^{\sqcup} M &:= (\{a,b\}\times[c,d]) \cup ([a,b]\times\{c\}) & \partial^{\lrcorner} M &:= (\{b\}\times[c,d]) \cup ([a,b]\times\{c\})
\end{aligned}$$

be the subsets of ∂M hinted by the pictorial superscript.

DEFINITION 1.50. The standard bicolored circle is $S^1 = \partial[0,1]^2$ with bicoloring $S^1_\circ = \partial^{\sqsubset}([0,\frac{1}{2}]\times[0,1])$ and $S^1_\bullet = \partial^{\sqsupset}([\frac{1}{2},1]\times[0,1])$. The upper half of this circle is $S^1_\top = \partial^{\sqcap}([0,1]\times[\frac{1}{2},1])$, and the standard involution is $(x,y) \mapsto (x, 1-y)$. The vacuum sector associated to the standard bicolored circle, its upper half, and its standard involution, is

$$H_0(D) := L^2(D(S^1_\top)).$$

It has left actions of $D(I)$ for every interval $I \subset S^1$ such that $(\frac{1}{2},0)$ and $(\frac{1}{2},1)$ are not both in I and $\{(\frac{1}{2},0),(\frac{1}{2},1)\} \cap \partial I = \emptyset$.

The fiber product $*$ of von Neumann algebras was studied in [**Tim08**]. It is an alternative to the fusion $\circledast$ of von Neumann algebras which remedies the formal shortcomings of the fusion operation (see Appendix B.IV). In view of this, one might rather have defined the composition of defects as

$$\text{(1.51)} \qquad (D *_{\mathcal{B}} E)(I) := \begin{cases} \mathcal{A}(I) & \text{for } I \in \mathsf{INT}_\circ \\ D(I^{++}) *_{\mathcal{B}(J)} E({}^{++}I) & \text{for } I \in \mathsf{INT}_{◑} \\ \mathcal{C}(I) & \text{for } I \in \mathsf{INT}_\bullet \end{cases}$$

where I^{++}, ${}^{++}I$, and J are as in (1.41). This is related to the previous definition (1.41) as follows. Let S^1 be the standard bicolored circle (see Definition 1.50), with upper and lower halves $S^1_\top$ and $S^1_\perp$.

LEMMA 1.52. *Let ${}_{\mathcal{A}}D_{\mathcal{B}}$ and ${}_{\mathcal{B}}E_{\mathcal{C}}$ be defects, with corresponding vacuum sectors $H := H_0(D)$ and $K := H_0(E)$. Viewed as algebras acting on $H \boxtimes_{\mathcal{B}} K$, we then have*

$$(D * E)(S^1_\top) = \big((D \circledast E)(S^1_\perp)\big)' \qquad \textit{and} \qquad (D * E)(S^1_\perp) = \big((D \circledast E)(S^1_\top)\big)'.$$

Proof. Using a graphical representation as in (1.42), we have:

$$\begin{aligned}(D * E)(S^1_\top) &= D(\ulcorner\!\rceil\!\!|) *_{\mathcal{B}(||)} E(|\!\ulcorner\!\urcorner) \\ &= \Big(D(\ulcorner\!\rceil\!\!|)' \otimes_{alg} E(|\!\ulcorner\!\urcorner)'\Big)' \qquad \text{(commutants taken on } H, K, \text{ and } H \boxtimes_{\mathcal{B}} K, \text{ respectively)} \\ &= \Big(D(\llcorner_{=}) \otimes_{alg} E(_{=}\lrcorner)\Big)' \\ &= \Big(D(\llcorner_{=}) \vee E(_{=}\lrcorner)\Big)' = ((D \circledast E)(S^1_\perp))'.\end{aligned}$$

where the third equality follows by Haag duality (Proposition 1.16). The second equation is similar. □

When the conformal net $\mathcal{B}$ has finite index (and conjecturally even without that restriction), the two definitions of fusion (1.41) and (1.51) actually agree:

Theorem 1.53. *Let ${}_{\mathcal{A}}D_{\mathcal{B}}$ and ${}_{\mathcal{B}}E_{\mathcal{C}}$ be defects. If $\mathcal{B}$ has finite index, then for every bicolored interval I the inclusion*

$$(D \circledast_{\mathcal{B}} E)(I) \hookrightarrow (D *_{\mathcal{B}} E)(I) \tag{1.54}$$

is an isomorphism.

Proof. If $I \in \mathsf{INT}_\circ$ or $I \in \mathsf{INT}_\bullet$, then there is nothing to show. Recall the Notation 1.49. For $I := \partial^\sqcap([0,1] \times [\frac{1}{2}, 1])$, with bicoloring $I_\circ := I_{x \le \frac{1}{2}}$ and $I_\bullet := I_{x \ge \frac{1}{2}}$, the equality $(D \circledast_{\mathcal{B}} E)(I) = (D *_{\mathcal{B}} E)(I)$ is the content of Corollary 5.9. The result follows as every genuinely bicolored interval is isomorphic to this interval. □

Using the above theorem, the associator

$$(D \circledast_{\mathcal{B}} E) \circledast_{\mathcal{C}} F \cong D \circledast_{\mathcal{B}} (E \circledast_{\mathcal{C}} F) \tag{1.55}$$

is then induced from the associator for the operation $*$ of fiber product of von Neumann algebras. If I is a genuinely bicolored interval, then evaluating the two sides of (1.55) on I yields

$$\Big(D(I^{++}) \underset{\mathcal{B}(J)}{*} E(K)\Big) \underset{\mathcal{C}(J')}{*} F({}^{++}I) \quad\text{and}\quad D(I^{++}) \underset{\mathcal{B}(J)}{*} \Big(E(K) \underset{\mathcal{C}(J')}{*} F({}^{++}I)\Big), \tag{1.56}$$

where

$$I^{++} = I_\circ \cup [0, \tfrac{3}{2}], \quad K = [-\tfrac{3}{2}, \tfrac{3}{2}], \quad {}^{++}I = [-\tfrac{3}{2}, 0] \cup I_\bullet, \quad J = J' = [0,1],$$

and the embeddings $I^{++} \hookleftarrow J \hookrightarrow K \hookleftarrow J' \hookrightarrow {}^{++}I$ are as in (1.41). The associator relating the two sides of (1.56) (see [**Tim08**, Prop. 9.2.8] for a construction) is the desired natural isomorphisms (1.55). The properties of (1.55) can then be summarized by saying that it provides a natural transformation

$$\mathsf{associator}\colon \mathsf{CN}_1 \times_{\mathsf{CN}_0^f} \mathsf{CN}_1 \times_{\mathsf{CN}_0^f} \mathsf{CN}_1 \Rrightarrow \mathsf{CN}_1 \tag{1.57}$$

that is an associator for the composition (1.48). This associator satisfies the pentagon identity by the corresponding pentagon identity for the operation $*$.

CHAPTER 2

Sectors

We will use the constant speed parametrization to identify the standard circle $\{z \in \mathbb{C} : |z| = 1\}$ with the boundary of the unit square $\partial[0,1]^2$. Under our identification, the points 1, i, -1, and $-i$ get mapped to $(1, \frac{1}{2})$, $(\frac{1}{2}, 1)$, $(0, \frac{1}{2})$, and $(\frac{1}{2}, 0)$, respectively.

Recall the Notation 1.49. Our standard circle $S^1 = \partial[0,1]^2$ has a standard bicoloring given by $S^1_\circ := \partial^\sqsubset([0, \frac{1}{2}] \times [0,1])$ and $S^1_\bullet := \partial^\sqsupset([\frac{1}{2}, 1] \times [0,1])$. Let INT_{S^1} be the poset of subintervals of S^1, and let $\mathsf{INT}_{S^1,\circ\bullet}$ be the sub-poset of intervals $I \subset S^1$ such that $(I \cap S^1_\circ, I \cap S^1_\bullet)$ is a bicoloring. Thus, an interval I is in $\mathsf{INT}_{S^1,\circ\bullet}$ if neither of the color-change points $(\frac{1}{2}, 0)$ and $(\frac{1}{2}, 1)$ are in its boundary, and if both $I \cap S^1_\circ$ and $I \cap S^1_\bullet$ are connected (possibly empty). We view $\mathsf{INT}_{S^1,\circ\bullet}$ as a (non-full) subcategory of $\mathsf{INT}_{\circ\bullet}$.

2.A. The category CN_2 of sectors

The elements of $\mathsf{INT}_{S^1,\circ\bullet}$ naturally fall into four classes:

$$
\begin{aligned}
\mathsf{INT}_{S^1,\circ} &:= \{I \subset S^1 \mid I \cap S^1_\bullet = \emptyset\} \\
\mathsf{INT}_{S^1,\bullet} &:= \{I \subset S^1 \mid I \cap S^1_\circ = \emptyset\} \\
\mathsf{INT}_{S^1,\top} &:= \{I \subset S^1 \mid (\tfrac{1}{2}, 1) \in I \setminus \partial I \text{ and } (\tfrac{1}{2}, 0) \notin I\} \\
\mathsf{INT}_{S^1,\bot} &:= \{I \subset S^1 \mid (\tfrac{1}{2}, 0) \in I \setminus \partial I \text{ and } (\tfrac{1}{2}, 1) \notin I\}.
\end{aligned}
\tag{2.1}
$$

Definition 2.2. Let $\mathcal{A}$ and $\mathcal{B}$ be conformal nets, and let ${}_{\mathcal{A}}D_{\mathcal{B}}$ and ${}_{\mathcal{A}}E_{\mathcal{B}}$ be $\mathcal{A}$-$\mathcal{B}$-defects. A *D-E-sector* is a Hilbert space H, equipped with homomorphisms

$$
\begin{aligned}
&\rho_I \colon \mathcal{A}(I) \to \mathbf{B}(H) &&\text{for} \quad I \in \mathsf{INT}_{S^1,\circ} \\
&\rho_I \colon D(I) \to \mathbf{B}(H) &&\text{for} \quad I \in \mathsf{INT}_{S^1,\top} \\
&\rho_I \colon \mathcal{B}(I) \to \mathbf{B}(H) &&\text{for} \quad I \in \mathsf{INT}_{S^1,\bullet} \\
&\rho_I \colon E(I) \to \mathbf{B}(H) &&\text{for} \quad I \in \mathsf{INT}_{S^1,\bot}
\end{aligned}
\tag{2.3}
$$

subject to the condition $\rho_I|_J = \rho_J$ whenever $J \subset I$. Moreover, if $I \in \mathsf{INT}_{S^1,\top}$ and $J \in \mathsf{INT}_{S^1,\bot}$ are intervals with disjoint interiors, then $\rho_I(D(I))$ and $\rho_J(E(J))$ are required to commute with each other. We write ${}_D H_E$ to indicate that $H = (H, \rho)$ is a D-E-sector. If $D = E$, then we say that H is a *D-sector*.

Pictorially we will draw a D-E-sector as follows:

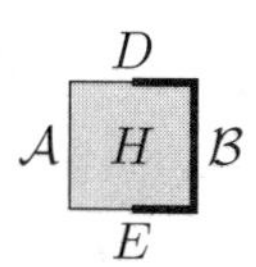

The thin line stands for the conformal net $\mathcal{A}$ and the thick line stands for $\mathcal{B}$.

REMARK 2.4. If $I \in \mathsf{INT}_{S^1,\top}$ and $J \in \mathsf{INT}_{S^1,\perp}$ are disjoint intervals, we do *not* require that the action of $D(I) \otimes_{alg} E(J)$ extends to an action of $D(I) \bar{\otimes} E(J)$.

Recall from Proposition 1.22 that if $\mathcal{A}$ and $\mathcal{B}$ are both equal to the trivial conformal net $\underline{\mathbb{C}}$, then a $\underline{\mathbb{C}}$-$\underline{\mathbb{C}}$-defect may be viewed simply as a von Neumann algebra. A D-E-sector between two such defects is given by a bimodule between the corresponding von Neumann algebras.

The following lemma is a straightforward analog of [**BDH13**, Lem. 1.9].

LEMMA 2.5. *Let S^1 be the standard bicolored circle, and let $\{I_i \subset S^1\}$ be a family of bicolored intervals whose interiors cover S^1. Suppose that we have actions*

$$\begin{aligned} \rho_i &\colon \mathcal{A}(I_i) \to \mathbf{B}(H), && \text{for} \quad I_i \in \mathsf{INT}_{S^1,\circ} \\ \rho_i &\colon D(I_i) \to \mathbf{B}(H), && \text{for} \quad I_i \in \mathsf{INT}_{S^1,\top} \\ \rho_i &\colon \mathcal{B}(I_i) \to \mathbf{B}(H), && \text{for} \quad I_i \in \mathsf{INT}_{S^1,\bullet} \\ \rho_i &\colon E(I_i) \to \mathbf{B}(H), && \text{for} \quad I_i \in \mathsf{INT}_{S^1,\perp}, \end{aligned}$$

subject to the following two conditions: 1. $\rho_i|_{I_i \cap I_j} = \rho_j|_{I_i \cap I_j}$, 2. if $J \subset I_i$ and $K \subset I_j$ are disjoint, then $\rho_i(\mathcal{A}(J))$ commutes with $\rho_j(\mathcal{A}(K))$. Then these actions endow H with the structure of a D-E-sector.

PROOF. Given an interval $J \subset S^1$, pick a diffeomorphism $\varphi \in \mathrm{Diff}_+(S^1)$ that is trivial in a neighborhood N of the two color changing points, and such that $\varphi(J) \subset I_{i_0}$ for some I_{i_0} in our cover. Write $\varphi = \varphi_n \circ \ldots \circ \varphi_1$ for diffeomorphisms φ_k that are trivial on N and whose supports lie in elements of the cover. Let u_k be unitaries implementing φ_k (Proposition 1.10). Upon identifying u_k with its image (under the relevant ρ_i) in $\mathbf{B}(H)$, we set

$$\rho_J(a) := u_1^* \ldots u_n^* \rho_{i_0}\big(\varphi(a)\big) u_n \ldots u_1. \tag{2.6}$$

Here we have used $\varphi(a)$ as an abbreviation for $\mathcal{A}(\varphi)(a)$, $D(\varphi)(a)$, $\mathcal{B}(\varphi)(a)$, or $E(\varphi)(a)$, depending on whether J is a white, top, black, or bottom interval. Finally, as in the proof of [**BDH13**, Lem. 1.9], one checks that $\rho_J|_K = \rho_{I_\ell}|_K$ for any sufficiently small interval $K \subset J \cap I_\ell$, and then uses strong additivity to conclude that $\rho_J|_{J \cap I_\ell} = \rho_{I_\ell}|_{J \cap I_\ell}$. □

As before let $S^1_\top = \partial^\sqcap([0,1]\times[\frac{1}{2},1])$ and $S^1_\perp = \partial^\sqcup([0,1]\times[0,\frac{1}{2}])$ be the upper and lower halves of the standard bicolored circle.

DEFINITION 2.7. Sectors form a category that we call CN_2. Its objects are quintuples $(\mathcal{A}, \mathcal{B}, D, E, H)$, where $\mathcal{A}$, $\mathcal{B}$ are semisimple conformal nets, D, E are $\mathcal{A}$-$\mathcal{B}$-defects, and H is a D-E-sector. A morphism from $(\mathcal{A}, \mathcal{B}, D, E, H)$ to $(\mathcal{A}', \mathcal{B}', D', E', H')$ consists of four compatible invertible natural transformations

$$\alpha\colon \mathcal{A} \to \mathcal{A}', \quad \beta\colon \mathcal{B} \to \mathcal{B}', \quad \delta\colon D \to D', \quad \varepsilon\colon E \to E',$$

along with a bounded linear map $h\colon H \to H'$ that is equivariant in the sense that

$$\begin{aligned} \rho'_I(\alpha(a)) \circ h &= h \circ \rho_I(a), & \rho'_I(\beta(b)) \circ h &= h \circ \rho_I(b), \\ \rho'_I(\delta(d)) \circ h &= h \circ \rho_I(d), & \rho'_I(\varepsilon(e)) \circ h &= h \circ \rho_I(e), \end{aligned}$$

for $a \in \mathcal{A}(I)$, $b \in \mathcal{B}(I)$, $d \in D(I)$, $e \in E(I)$, and $I \in \mathsf{INT}_{S^1,\circ}$, $\mathsf{INT}_{S^1,\bullet}$, $\mathsf{INT}_{S^1,\top}$, $\mathsf{INT}_{S^1,\perp}$ respectively.

There is also a symmetric monoidal structure on CN_2 given by objectwise spatial tensor product for the functors $\mathcal{A}$, $\mathcal{B}$, D, E, and by tensor product of Hilbert spaces.

The category CN_2 is equipped with two forgetful functors

$$\mathsf{source_v}\colon \mathsf{CN}_2 \to \mathsf{CN}_1 \qquad \mathsf{target_v}\colon \mathsf{CN}_2 \to \mathsf{CN}_1$$

called 'vertical source' and 'vertical target', given by $\mathsf{source_v}(\mathcal{A},\mathcal{B},D,E,H) = (\mathcal{A},\mathcal{B},D)$ and $\mathsf{target_v}(\mathcal{A},\mathcal{B},D,E,H) = (\mathcal{A},\mathcal{B},E)$. They satisfy

$$\mathsf{source} \circ \mathsf{source_v} = \mathsf{source} \circ \mathsf{target_v} \quad \text{and} \quad \mathsf{target} \circ \mathsf{source_v} = \mathsf{target} \circ \mathsf{target_v}.$$

Provided we restrict to the subcategory $\mathsf{CN}_1^f \subset \mathsf{CN}_1$ whose objects are semisimple defects between semisimple conformal nets and whose morphisms are finite natural transformations (another option is to allow all defects between semisimple conformal nets but restrict the morphisms to be only the isomorphisms), there is also a 'vertical identity' functor

$$\mathsf{identity_v}\colon \mathsf{CN}_1^f \to \mathsf{CN}_2 \tag{2.8}$$

that sends an $\mathcal{A}$-$\mathcal{B}$-defect D to the object $(\mathcal{A},\mathcal{B},D,D,H_0(D))$ of CN_2. Here, the vacuum sector $H_0(D) := L^2(D(S^1_\top)) = L^2(D(\ulcorner\!\urcorner))$ is as described in Definition 1.50. We represent it pictorially as follows:

(2.9) $\mathsf{identity_v}({}_{\mathcal{A}}D_{\mathcal{B}}) =$

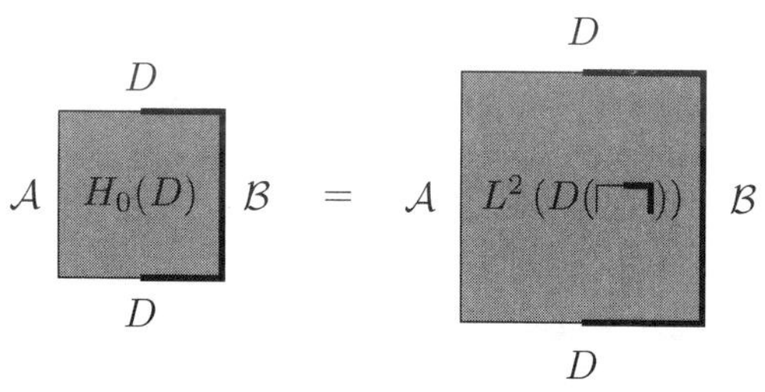

We reserve this darker shading of the above squares for vacuum sectors. Note that it is essential to restrict to the subcategory $\mathsf{CN}_1^f \subset \mathsf{CN}_1$ because the L^2-space construction is only functorial with respect to finite homomorphisms of von Neumann algebras [**BDH11**] (see also [**BDH11**, Conj. 6.17]).

REMARK 2.10. We will see later, in Warning 6.8, that we will have to further restrict our morphisms, and only allow natural *isomorphisms* between defects (even if the defects are semisimple). This will render otiose the subtleties related to [**BDH11**, Conj. 6.17]; in particular, there is no need to restrict to semisimple defects.

2.B. Horizontal fusion

Consider the translate $S^1_+ := \partial([1,2]\times[0,1]) \subset \mathbb{R}^2$ of the standard circle $S^1 = \partial[0,1]^2$, and let $\mathsf{INT}_{S^1_+,\circ}$, $\mathsf{INT}_{S^1_+,\bullet}$, $\mathsf{INT}_{S^1_+,\top}$, $\mathsf{INT}_{S^1_+,\perp}$ be the obvious analogs of (2.1). Given conformal nets $\mathcal{A}$, $\mathcal{B}$, $\mathcal{C}$, defects ${}_{\mathcal{A}}D_{\mathcal{B}}$, ${}_{\mathcal{A}}E_{\mathcal{B}}$, ${}_{\mathcal{B}}F_{\mathcal{C}}$, ${}_{\mathcal{B}}G_{\mathcal{C}}$, and sectors ${}_DH_E$, ${}_FK_G$, let us replace the structure maps (2.3) of K by actions

$$\rho_I : \mathcal{B}(I) \to \mathbf{B}(K) \text{ for } I \in \mathsf{INT}_{S^1_+,\circ} \qquad \rho_I : \mathcal{C}(I) \to \mathbf{B}(K) \text{ for } I \in \mathsf{INT}_{S^1_+,\bullet}$$

$$\rho_I : F(I) \to \mathbf{B}(K) \text{ for } I \in \mathsf{INT}_{S^1_+,\top} \qquad \rho_I : G(I) \to \mathbf{B}(K) \text{ for } I \in \mathsf{INT}_{S^1_+,\perp}$$

by precomposing with the translation. Let J be $\{1\} \times [0,1] = S^1 \cap S^1_+$, with the orientation inherited from S^1_+. The algebra $\mathcal{B}(J)$ has actions of opposite variance on H and on K, so it makes sense to take the Connes fusion

$$H \boxtimes_{\mathcal{B}} K := H \boxtimes_{\mathcal{B}(J)} K.$$

We now show that $H \boxtimes_{\mathcal{B}} K$ is a $(D \circledast_{\mathcal{B}} F)$-$(E \circledast_{\mathcal{B}} G)$-sector. Given $I \in \mathsf{INT}_{S^1,\circ\bullet}$, let I_+ be the subinterval of $\partial([0,2] \times [0,1])$ given by

$$\begin{aligned} I_+ &:= I && \text{if } I \in \mathsf{INT}_{S^1,\circ}, \\ I_+ &:= I + (1,0) && \text{if } I \in \mathsf{INT}_{S^1,\bullet}, \\ I_+ &:= I_\circ \cup \left([\tfrac{1}{2}, \tfrac{3}{2}] \times \{1\}\right) \cup \left(I_\bullet + (1,0)\right) && \text{if } I \in \mathsf{INT}_{S^1,\top}, \\ I_+ &:= I_\bullet \cup \left([\tfrac{1}{2}, \tfrac{3}{2}] \times \{0\}\right) \cup \left(I_\bullet + (1,0)\right) && \text{if } I \in \mathsf{INT}_{S^1,\bot}. \end{aligned}$$

For $I \in \mathsf{INT}_{S^1,\circ}$ and $I \in \mathsf{INT}_{S^1,\bullet}$, the structure maps (2.3) are given by the obvious actions of $\mathcal{A}(I_+)$ and $\mathcal{C}(I_+)$ on the Hilbert space $H \boxtimes_{\mathcal{B}} K$. For an interval $I \in \mathsf{INT}_{S^1,\top}$ or $I \in \mathsf{INT}_{S^1,\bot}$, the algebras that act on $H \boxtimes_{\mathcal{B}(J)} K$ are given by

$$(2.11) \qquad \begin{aligned} &D\big((I_+ \cap S^1) \cup J\big) \circledast_{\mathcal{B}(J)} F\big(J \cup (I_+ \cap S^1_+)\big) \\ \text{and} \qquad &E\big((I_+ \cap S^1) \cup J\big) \circledast_{\mathcal{B}(J)} G\big(J \cup (I_+ \cap S^1_+)\big) \end{aligned}$$

respectively—see Appendix B.IV. Upon identifying the intervals $(I_+ \cap S^1) \cup J$ and $J \cup (I_+ \cap S^1_+)$ of (2.11) with the intervals I^{++} and ${}^{++}I$ of (1.41), we see that the algebras (2.11) are equal to $\big(D \circledast_{\mathcal{B}} F\big)(I)$ and $\big(E \circledast_{\mathcal{B}} G\big)(I)$, respectively. We can now define the functor of horizontal fusion

$$(2.12) \qquad \mathsf{fusion_h} : \mathsf{CN}_2 \times_{\mathsf{CN}_0^f} \mathsf{CN}_2 \to \mathsf{CN}_2$$

by $\mathsf{fusion_h}\big((\mathcal{A}, \mathcal{B}, D, E, H), (\mathcal{B}, \mathcal{C}, F, G, K)\big) := \big(\mathcal{A},\, \mathcal{C},\, D \circledast_{\mathcal{B}} F,\, E \circledast_{\mathcal{B}} G,\, H \boxtimes_{\mathcal{B}} K\big)$. Here, as in (1.48), $\mathsf{CN}_2 \times_{\mathsf{CN}_0^f} \mathsf{CN}_2$ is a shorthand notation for $\mathsf{CN}_2 \times_{\mathsf{CN}_0} \mathsf{CN}_0^f \times_{\mathsf{CN}_0} \mathsf{CN}_2$, and the relevant maps $\mathsf{CN}_2 \to \mathsf{CN}_0$ are $\mathsf{target} \circ \mathsf{source_v}$ and $\mathsf{source} \circ \mathsf{source_v}$, respectively.

Pictorially, we understand the functor $\mathsf{fusion_h}$ as the operation of gluing two squares along a common edge as follows:

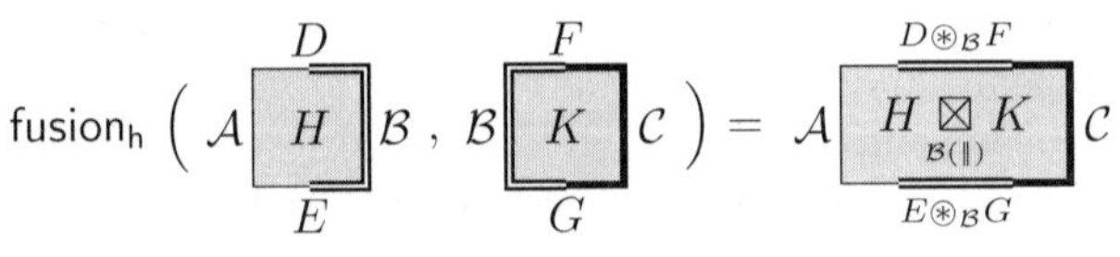

The associator for $\mathsf{fusion_h}$ is induced by the usual associator for Connes fusion. It consists of a natural transformation

$$(2.13) \qquad \mathsf{associator_h} : \mathsf{CN}_2 \times_{\mathsf{CN}_0^f} \mathsf{CN}_2 \times_{\mathsf{CN}_0^f} \mathsf{CN}_2 \Rrightarrow \mathsf{CN}_2,$$

and satisfies the pentagon identity.

2.C. Vertical fusion

Unlike horizontal fusion, vertical fusion is not the operation of gluing two squares along a common edge. Rather, it consists of gluing those two squares along half of their boundary:

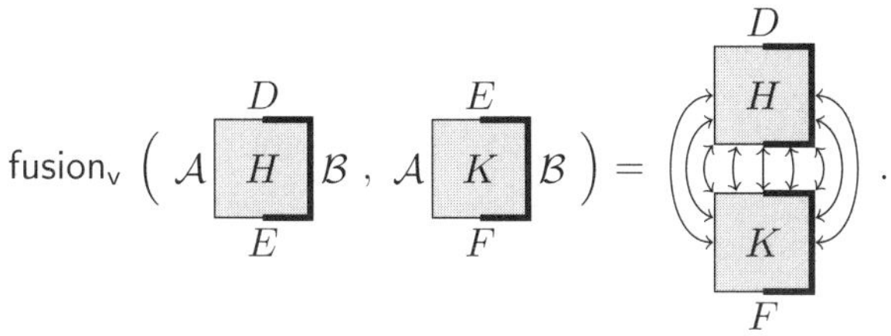

A sector is called *dualizable* if it has a dual with respect to the operation of vertical fusion; equivalently:

DEFINITION 2.14. A D-E-sector H between semisimple defects is called dualizable if it is dualizable (Appendix B.VI) as an $S^1_\top(D)$–$S^1_\bot(E)$-bimodule.

We now describe in detail the functor $\mathsf{fusion}_\mathsf{v}$ of vertical fusion. Given conformal nets $\mathcal{A}$, $\mathcal{B}$, defects ${}_\mathcal{A}D_\mathcal{B}$, ${}_\mathcal{A}E_\mathcal{B}$, ${}_\mathcal{A}F_\mathcal{B}$, and sectors ${}_DH_E$, ${}_EK_F$, we want to construct a D-F-sector $H \boxtimes_E K$. Let $S^1_\top = \partial^\sqcap([0,1]\times[\frac{1}{2},1])$ and $S^1_\bot = \partial^\sqcup([0,1]\times[0,\frac{1}{2}])$ be the top and bottom halves of our standard circle $\partial[0,1]^2$, and let $j : S^1_\top \xrightarrow{\sim} S^1_\bot$ be the reflection map along the horizontal symmetry axis. The algebra $E(S^1_\top)$ has two actions

$$E(S^1_\top)^{op} \xrightarrow{E(j)} E(S^1_\bot) \to \mathbf{B}(H)$$
$$E(S^1_\top) \longrightarrow \mathbf{B}(K)$$

of opposite variance on H and K, and so it makes sense to take the Connes fusion

$$H \boxtimes_E K := H \boxtimes_{E(S^1_\top)} K.$$

To see that $H \boxtimes_E K$ is a D-F-sector, we have to show that the algebras $\mathcal{A}(I)$, $\mathcal{B}(I)$, $D(I)$, and $F(I)$ act on it for $I \in \mathsf{INT}_{S^1,\circ}$, $\mathsf{INT}_{S^1,\bullet}$, $\mathsf{INT}_{S^1,\top}$, $\mathsf{INT}_{S^1,\bot}$, respectively.

We first treat the case $I \in \mathsf{INT}_{S^1,\circ}$. If I is contained in $S^1_\top$ (or $S^1_\bot$), then the action of $\mathcal{A}(I)$ on $H \boxtimes_E K$ is induced by its action on H (or K). If I contains the point $(0,\frac{1}{2})$ in its interior, then the algebra

$$E(I \cup S^1_\bot) \underset{E(S^1_\top)}{\circledast} E(I \cup S^1_\top) \qquad (2.15)$$

acts on $H \boxtimes_{E(S^1_\top)} K$, where the homomorphism $E(S^1_\top) \to E(I \cup S^1_\bot)^{op}$ implicit in (2.15) is given by $E(j)$. We observe, as follows, that there is a canonical homomorphism (typically not an isomorphism) from $\mathcal{A}(I)$ to the algebra (2.15). In the definition of that fusion product, we are free to chose any faithful $E(I \cup S^1_\bot)$-module and any faithful $E(I \cup S^1_\top)$-module (see Appendix B.IV): let us take both of them to be the vacuum $H_0(E)$. Then, by definition, the algebra (2.15) is generated on

$$H_0(E) \boxtimes_{E(S^1_\top)} H_0(E) \cong H_0(E)$$

by $E(I \cup S^1_\bot) \cap E(S^1_\bot)'$ and $E(I \cup S^1_\top) \cap E(S^1_\top)'$. By the strong additivity, vacuum, and locality axioms, we have natural homomorphisms (the first one is an isomorphism when E is a faithful defect):

$$\begin{aligned}\mathcal{A}(I) &\to E(I \cap S^1_\top) \vee E(I \cap S^1_\bot)\\ &\to \big(E(I \cup S^1_\bot) \cap E(S^1_\bot)'\big) \vee \big(E(I \cup S^1_\top) \cap E(S^1_\top)'\big)\\ &= E(I \cup S^1_\bot) \circledast_{E(S^1_\top)} E(I \cup S^1_\top).\end{aligned}$$

Composing this composite with the action of (2.15) on $H \boxtimes_E K$ gives our desired action of $\mathcal{A}(I)$.

By the same argument, we also have actions of $\mathcal{B}(I)$ on $H \boxtimes_E K$ for $I \in \mathsf{INT}_{S^1,\bullet}$. Furthermore, there are actions of $D(S^1_\top)$ and $F(S^1_\bot)$ on $H \boxtimes_E K$ coming from their respective actions on H and on K. We can therefore apply Lemma 2.5 to all the actions constructed so far, and conclude that $H \boxtimes_E K$ is a D-F-sector.

One might expect vertical fusion to be a functor $\mathsf{CN}_2 \times_{\mathsf{CN}_1} \mathsf{CN}_2 \to \mathsf{CN}_2$. However, just like the vertical identity (2.8) which is only a functor on the smaller category CN_1^f, and the horizontal fusion which is only a functor on the restricted product $\mathsf{CN}_2 \times_{\mathsf{CN}_0^f} \mathsf{CN}_2$, so too vertical fusion only gives a functor on the restricted product:

$$\mathsf{fusion_v} : \mathsf{CN}_2 \times_{\mathsf{CN}_1^f} \mathsf{CN}_2 \to \mathsf{CN}_2 \tag{2.16}$$

$$\mathsf{fusion_v}\big((\mathcal{A}, \mathcal{B}, D, E, H), (\mathcal{A}, \mathcal{B}, E, F, K)\big) := \big(\mathcal{A},\, \mathcal{B},\, D, F,\, H \boxtimes_E K\big).$$

The restriction is necessary to ensure the Connes fusion $H \boxtimes_E K$ is functorial with respect to the relevant natural transformations of the defect E [**BDH11**].

The associator for vertical fusion

$$\mathsf{associator_v} : \mathsf{CN}_2 \times_{\mathsf{CN}_1^f} \mathsf{CN}_2 \times_{\mathsf{CN}_1^f} \mathsf{CN}_2 \Rrightarrow \mathsf{CN}_2 \tag{2.17}$$

comes from the associator of Connes fusion and satisfies the pentagon identity. There are also 'top' and 'bottom' identity natural transformations,

$$\mathsf{unitor_t} : \mathsf{CN}_2 \Rrightarrow \mathsf{CN}_2, \qquad \mathsf{unitor_b} : \mathsf{CN}_2 \Rrightarrow \mathsf{CN}_2 \tag{2.18}$$

that describe the way $\mathsf{fusion_v}$ and $\mathsf{identity_v}$ interact. Given a sector ${}_DH_E$, they provide natural isomorphisms

$${}_DH_0(D) \boxtimes_D H_E \cong {}_DH_E \qquad \text{and} \qquad {}_DH \boxtimes_E H_0(E)_E \cong {}_DH_E \tag{2.19}$$

subject to the usual triangle axioms. Strictly speaking, the source functor of $\mathsf{unitor_t}$ is only defined on the subcategory $\mathsf{CN}_1^f \times_{\mathsf{CN}_1} \mathsf{CN}_2$ of CN_2, and so the transformation $\mathsf{unitor_t}$ itself is only defined on that subcategory. Similarly, $\mathsf{unitor_b}$ is only defined on the subcategory $\mathsf{CN}_2 \times_{\mathsf{CN}_1} \mathsf{CN}_1^f$.

CHAPTER 3

Properties of the composition of defects

3.A. Left and right units

Units are a subtle business. One might guess that the left unit is a natural isomorphism $\mathsf{CN}_1 \Downarrow \mathsf{CN}_1$ whose source is the functor $\mathsf{composition} \circ ((\mathsf{identity} \circ \mathsf{source}) \times \mathrm{id}_{\mathsf{CN}_1})$ and whose target is the identity functor. (Here $\mathrm{id}_{\mathsf{CN}_1} : \mathsf{CN}_1 \to \mathsf{CN}_1$ is the identity functor and $\mathsf{identity} : \mathsf{CN}_0 \to \mathsf{CN}_1$ takes a net to the identity defect, as in (1.39).) But, unfortunately, in general there is no such natural isomorphism. Instead, we have the following 'weaker' piece of data: a functor

$$\mathsf{unitor}_{\mathsf{tl}} : \mathsf{CN}_1^f \to \mathsf{CN}_2$$

('tl' stands for top left) with the property that

$$\mathsf{source}_{\mathsf{v}} \circ \mathsf{unitor}_{\mathsf{tl}} = \mathsf{composition} \circ ((\mathsf{identity} \circ \mathsf{source}) \times \mathrm{id}_{\mathsf{CN}_1^f})$$
$$\text{and} \qquad \mathsf{target}_{\mathsf{v}} \circ \mathsf{unitor}_{\mathsf{tl}} = \mathrm{id}_{\mathsf{CN}_1^f}\,.$$

This functor takes values in sectors that are invertible with respect to vertical fusion. Its construction is based on the following lemma:

LEMMA 3.1. *Let ${}_{\mathcal{A}}D_{\mathcal{B}}$ be a defect. Then $1 \circledast D := \mathsf{identity}(\mathcal{A}) \circledast_{\mathcal{A}} D$ is given on genuinely bicolored intervals I by*

$$\big(1 \circledast D\big)(I) = D({}^{<}I)$$

where ${}^{<}I := I_\circ \cup [0,1] \cup I_\bullet$ with bicoloring ${}^{<}I_\circ := I_\circ \cup [0,1]$ and ${}^{<}I_\bullet := I_\bullet$.

Similarly, on genuinely bicolored intervals we have

$$(D \circledast 1)(I) = D(I^{>})$$

where $I^{>} := I_\circ \cup [0,1] \cup I_\bullet$ with bicoloring $I^{>}_\circ := I_\circ$ and $I^{>}_\bullet := [0,1] \cup I_\bullet$.

PROOF. We prove the first statement; the second one is entirely similar. Consider the intervals $K := \{\frac{1}{2}\} \times [0,1]$, $J := I_\circ \cup [0,\frac{1}{2}] \times \{1\}$, $J^+ := J \cup K$, ${}^{+}I := ([\frac{1}{2},1] \times \{1\}) \cup I_\bullet$, and ${}^{++}I := {}^{+}I \cup K$. These intervals are bicolored by ${}^{+}I_\bullet = {}^{++}I_\bullet = I_\bullet$ and $K_\bullet = J_\bullet = J^+_\bullet = \emptyset$.

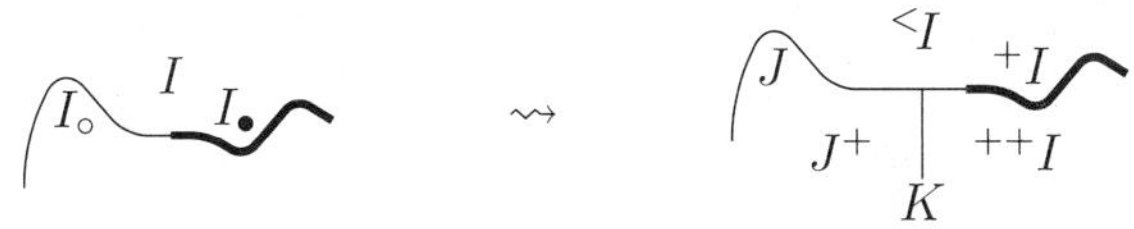

Extend the map $[0,\frac{1}{2}] \times \{1\} \to K : (t,1) \mapsto (\frac{1}{2}, t+\frac{1}{2})$ to an embedding $f : J \to K$ so that $K \setminus f(J)$ is non-empty. Using $\mathcal{A}(f)$, we can then equip $L^2(\mathcal{A}(K))$ with a left action of $\mathcal{A}(J)$. Combining this left action with the natural right action of $\mathcal{A}(K)$, we get a faithful action of $\mathcal{A}(J) \otimes_{alg} \mathcal{A}(K)^{op}$ on $L^2(\mathcal{A}(K))$, which extends to $\mathcal{A}(J^+)$ by the vacuum sector axiom for conformal nets (see Appendix C.I). Pick a faithful $D({}^{++}I)$-module H. By definition,

$$\big(1 \circledast D\big)(I) \;=\; \mathcal{A}(J^+) \circledast_{\mathcal{A}(K)} D({}^{++}I)$$

is the von Neumann algebra generated by $\mathcal{A}(J)$ and $D({}^{+}I)$ on the Hilbert space $L^2(\mathcal{A}(K)) \boxtimes_{\mathcal{A}(K)} H \cong H$. This algebra is equal to $D(J \cup {}^{+}I) = D({}^{<}I)$ by strong additivity. □

Recall that our standard circle S^1 is the square $\partial[0,1]^2$. Let

$$S^1_{\ulcorner} := \partial^{\ulcorner}\big([0,\tfrac{1}{2}]\times[\tfrac{1}{2},1]\big), \qquad S^1_{\urcorner} := \partial^{\urcorner}\big([\tfrac{1}{2},1]\times[\tfrac{1}{2},1]\big),$$
$$S^1_{\llcorner} := \partial^{\llcorner}\big([0,\tfrac{1}{2}]\times[0,\tfrac{1}{2}]\big), \qquad S^1_{\lrcorner} := \partial^{\lrcorner}\big([\tfrac{1}{2},1]\times[0,\tfrac{1}{2}]\big)$$

be the four "quarter circles". Let us also pick, once and for all, a diffeomorphism $\phi_{\ulcorner} : S^1_{\ulcorner} \cup [0,1] \to S^1_{\ulcorner}$ (here $(\frac{1}{2},1) \in S^1_{\ulcorner}$ is glued to $0 \in [0,1]$) whose derivative is equal to one in a neighborhood of the boundary. The three mirror images of $\phi_{\ulcorner}$ are called $\phi_{\urcorner} : [0,1] \cup S^1_{\urcorner} \to S^1_{\urcorner}$, $\phi_{\llcorner} : S^1_{\llcorner} \cup [0,1] \to S^1_{\llcorner}$, and $\phi_{\lrcorner} : [0,1] \cup S^1_{\lrcorner} \to S^1_{\lrcorner}$:

$\phi_{\ulcorner}$: 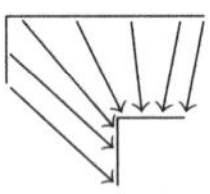 $\phi_{\urcorner}$: 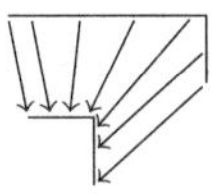 $\phi_{\llcorner}$: 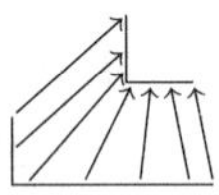 $\phi_{\lrcorner}$: 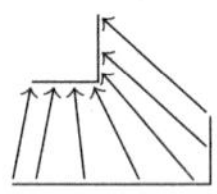

We are now ready to define the functor

$$\text{(3.2)} \qquad \mathsf{unitor_{tl}} \;:\; \mathsf{CN}_1 \to \mathsf{CN}_2.$$

It assigns to every $\mathcal{A}$-$\mathcal{B}$-defect D an invertible $(1 \circledast D)$-D-sector. As a Hilbert space, $\mathsf{unitor_{tl}}(D)$ is simply the vacuum sector $H_0(D)$. Let $\overline{S}$ be the bicolored circle with white half $\overline{S}_\circ := S^1_\circ \cup_{(\frac{1}{2},1)} [0,1]$ and black half $\overline{S}_\bullet := S^1_\bullet$. One should imagine $\overline{S}$ as being the standard bicolored circle S^1, to which an extra white interval $[0,1]$ has been inserted at the top—see (3.4). In view of Lemma 3.1, a $(1 \circledast D)$-D-sector is the same thing as a Hilbert space H equipped with compatible actions of $D(I)$ for every bicolored interval $I \subset \overline{S}$. Let $\hat{\phi}_{\ulcorner} : \overline{S} \to S^1$ be the diffeomorphism given by $\phi_{\ulcorner}$ on $S^1_{\ulcorner} \cup [0,1]$, and by the identity on the complement. The $(1 \circledast D)$-D-sector structure on $H_0(D) = \mathsf{unitor_{tl}}(D)$ is given by letting $D(I)$ act by the composition of $D(\hat{\phi}_{\ulcorner}) : D(I) \to D(\hat{\phi}_{\ulcorner}(I))$ with the natural action of $D(\hat{\phi}_{\ulcorner}(I))$ on $H_0(D)$.

We also have functors

$$\text{(3.3)} \qquad \begin{aligned} \mathsf{unitor_{tr}} \;&:\; \mathsf{CN}_1^f \to \mathsf{CN}_2 \\ \mathsf{unitor_{bl}} \;&:\; \mathsf{CN}_1^f \to \mathsf{CN}_2 \\ \mathsf{unitor_{br}} \;&:\; \mathsf{CN}_1^f \to \mathsf{CN}_2 \end{aligned}$$

that are defined in a similar fashion. The underlying Hilbert spaces of $\mathsf{unitor_{tr}}(D)$, $\mathsf{unitor_{bl}}(D)$, and $\mathsf{unitor_{br}}(D)$ are all $H_0(D)$, and they are equipped with the structures of $(D \circledast 1)$-D-sector, D-$(1 \circledast D)$-sector, and D-$(D \circledast 1)$-sector, respectively. Let $\overline{S}$, $\underline{S}$, and $\underline{S}$ be the bicolored circles given by $\overline{S}_\circ := S^1_\circ$, $\overline{S}_\bullet := [0,1] \cup_{(\frac{1}{2},1)} S^1_\bullet$, $\underline{S}_\circ := S^1_\circ \cup_{(\frac{1}{2},0)} [0,1]$, $\underline{S}_\bullet := S^1_\bullet$, and $\underline{S}_\circ := S^1_\circ$, $\underline{S}_\bullet := [0,1] \cup_{(\frac{1}{2},0)} S^1_\bullet$:

(3.4) $\overline{S}$ =  $\overline{S}$ =  $\underline{S}$ =  $\underline{S}$ = .

By Lemma 3.1, a $(D \circledast 1)$-D-sector structure on a Hilbert space is the same thing as a collection of compatible actions of the algebras $D(I)$ for every bicolored interval

$I \subset \overline{S}$. Similarly, being a D-$(1 \circledast D)$-sector means that there are compatible actions of $D(I)$ for every $I \subset \underline{S}$, and being D-$(D \circledast 1)$-sector means that there are compatible actions of $D(I)$ for every $I \subset \underline{S}$. We equip $H_0(D)$ with the above structures by the appropriate analogs $\hat{\phi}_{\urcorner} : \overline{S} \to S^1$, $\hat{\phi}_{\llcorner} : \underline{S} \to S^1$, $\hat{\phi}_{\lrcorner} : \underline{S} \to S^1$ of $\hat{\phi}_{\ulcorner}$, defined as $\phi_{\urcorner}$, $\phi_{\llcorner}$, $\phi_{\lrcorner}$ on the relevant subintervals, and as the identity on the rest.

EXAMPLE 3.5. Given a non-trivial conformal net $\mathcal{A}$, the identity defect $1_{\mathcal{A}} := \mathsf{identity}(\mathcal{A}) : I \mapsto \mathcal{A}(I)$ is not isomorphic to $1_{\mathcal{A}} \circledast_{\mathcal{A}} 1_{\mathcal{A}}$. The defect $1_{\mathcal{A}} \circledast_{\mathcal{A}} 1_{\mathcal{A}}$ is the weak identity for $\mathcal{A}$ discussed in Remark 1.40. It maps a genuinely bicolored interval I to $\mathcal{A}(I_\circ \cup [0,1] \cup I_\bullet)$. As a way of distinguishing those two defects, note that the intersection

$$\bigcap_{\substack{J \subset I,\ \text{genuinely} \\ \text{bicolored}}} (1_{\mathcal{A}} \circledast_{\mathcal{A}} 1_{\mathcal{A}})(J) = \mathcal{A}([0,1])$$

is non-trivial, which is not the case if $1_{\mathcal{A}} \circledast_{\mathcal{A}} 1_{\mathcal{A}}$ is replaced by $1_{\mathcal{A}}$ in the above expression.

The invertible sector between $1_{\mathcal{A}}$ and $1_{\mathcal{A}} \circledast_{\mathcal{A}} 1_{\mathcal{A}}$ is the vacuum module of $\mathcal{A}$ associated to the "circle"  constructed by inserting a copy of $[0,1]$ at the point $(\frac{1}{2}, 1) \in \partial[0,1]^2$.

3.B. Semisimplicity of the composite defect

Given two semisimple defects (finite direct sums of irreducible defects), we can ask whether their fusion is again a semisimple defect. From now on, we always assume that our conformal nets are irreducible. The purpose of this section is to prove the following theorem:

THEOREM 3.6. *Let ${}_{\mathcal{A}}D_{\mathcal{B}}$ and ${}_{\mathcal{B}}E_{\mathcal{C}}$ be semisimple defects. If the conformal net $\mathcal{B}$ has finite index, then for any genuinely bicolored interval I the algebra $(D \circledast_{\mathcal{B}} E)(I)$ has finite dimensional center.*

In light of Theorem 1.44 (whose proof, note well, depends on Theorem 3.6, via Corollary 4.16 and Proposition 4.18), we can rephrase this result as follows:

COROLLARY 3.7. *The fusion of two semisimple defects ${}_{\mathcal{A}}D_{\mathcal{B}}$ and ${}_{\mathcal{B}}E_{\mathcal{C}}$, over a finite index conformal net $\mathcal{B}$, is a semisimple $\mathcal{A}$-$\mathcal{C}$-defect.*

Detecting semisimplicity. We begin with a few lemmas.

LEMMA 3.8. *Let A, B be von Neumann algebras and let H be a faithful A–B-bimodule. If the algebra of A–B-bimodule endomorphisms of H is finite-dimensional, then A is a finite direct sums of factors.*

PROOF. The center of A acts faithfully by A–B-bimodule endomorphisms. It is therefore finite-dimensional. □

From now on, we fix a faithful defect ${}_{\mathcal{A}}D_{\mathcal{B}}$, and denote its vacuum sector H_0. Recall that our standard circle is $S^1 := \partial[0,1]^2$, and that its top and bottom halves are denoted $S^1_\top$ and $S^1_\bot$.

NOTATION 3.9. Given an interval $I \subset S^1$ that contains the two color-change points $(\frac{1}{2}, 0)$ and $(\frac{1}{2}, 1)$ in its interior, we define an algebra $\tilde{D}(I) \subset \mathbf{B}(H_0)$ as follows. It is the algebra generated by $D(I_1)$ and $D(I_2)$, where I_1 and I_2 are any two intervals covering I with the property that $(\frac{1}{2},1) \notin I_1$ and $(\frac{1}{2},0) \notin I_2$. By strong additivity, the algebra $\tilde{D}(I)$ does not depend on the choice of covering.

LEMMA 3.10. *Let $I \subset S^1$ be an interval containing both color-change points in its interior. If D is an irreducible defect, then $\tilde{D}(I)$ is a factor.*

PROOF. Let I' be the closure of $S^1 \backslash I$. The center of $\tilde{D}(I)$ commutes with both $D(I)$ and $D(I')$. Since $D(S^1_\top \cap I)$ and $D(S^1_\top \cap I')$ generate $D(S^1_\top)$, $Z(\tilde{D}(I))$ commutes with $D(S^1_\top)$. Similarly, $Z(\tilde{D}(I))$ commutes with $D(S^1_\perp)$. Therefore, $Z(\tilde{D}(I))$ acts on H_0 by $D(S^1_\top)$–$D(S^1_\perp)^{op}$-bimodule automorphisms. If $\tilde{D}(I)$ was not a factor, that action could be used to construct a non-trivial direct sum decomposition of $H_0 = L^2(D(S^1_\top))$, contradicting the irreducibility of D. □

Finiteness of the defect vacuum as a 4-interval bimodule (*splitting* ).

NOTATION 3.11. Let $S^1 = I_1 \cup I_2 \cup I_3 \cup I_4$ be a partitioning of the standard bicolored circle into four intervals so that I_1 and I_4 are genuinely bicolored, and so that each intersection $I_i \cap I_{i+1}$ (cyclic numbering) is a single point. For such a partition, we let $\hat{D}(I_1 \cup I_3)$ denote the commutant of $D(I_2 \cup I_4) = D(I_2) \mathbin{\bar{\otimes}} D(I_4)$ acting on the vacuum sector $H_0(D)$.

Similarly, if $\mathcal{A}$ is a conformal net and $S^1 = I_1 \cup I_2 \cup I_3 \cup I_4$ is a partitioning of the standard (not bicolored) circle, we let $\hat{\mathcal{A}}(I_1 \cup I_3)$ denote the commutant of $\mathcal{A}(I_2 \cup I_4) = \mathcal{A}(I_2) \mathbin{\bar{\otimes}} \mathcal{A}(I_4)$ on the vacuum sector $H_0(\mathcal{A})$.

Note that the choice of ambient circle does not affect the resulting algebras $\hat{D}(I_1 \cup I_3)$ and $\hat{\mathcal{A}}(I_1 \cup I_3)$: they only depend (up to canonical isomorphism) on the intervals I_1 and I_3, and on the bicoloring of those intervals.

LEMMA 3.12. *Let I_1, I_2, I_3, I_4 be as in Notation 3.11. Assume furthermore that I_2 and I_3 are white. Write $I_1 = J_1 \cup J_2$, with J_1 genuinely bicolored, J_2 white, and $J_1 \cap J_2$ a single point:*

(3.13)

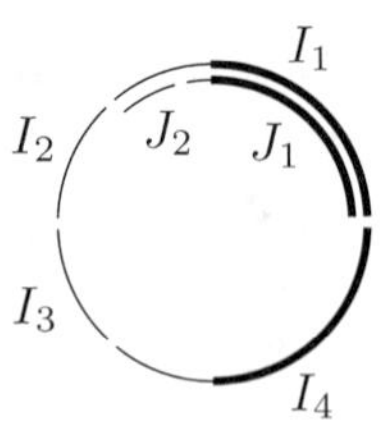

Then there is a natural action of the algebra $\hat{\mathcal{A}}(J_2 \cup I_3)$ on the vacuum sector $H_0(D)$, and we have $\hat{D}(I_1 \cup I_3) = D(J_1) \vee \hat{\mathcal{A}}(J_2 \cup I_3)$.

PROOF. We assume that D is faithful; otherwise $D = 0$ (since $\mathcal{A}$ and $\mathcal{B}$ are irreducible) and there is nothing to show. By Haag duality for $\mathcal{A}$, the algebra $\hat{\mathcal{A}}(J_2 \cup I_3)$ is $\mathcal{A}(J_2 \cup I_2 \cup I_3) \cap \mathcal{A}(I_2)'$, where the commutant is taken in the action on $H_0(\mathcal{A})$. The algebra $\mathcal{A}(J_2 \cup I_2 \cup I_3)$ also naturally acts on $H_0(D)$, and therefore so does $\hat{\mathcal{A}}(J_2 \cup I_3)$. Because of the faithfulness assumption, the algebra $\hat{\mathcal{A}}(J_2 \cup I_3)$ may equally well be expressed as $\mathcal{A}(J_2 \cup I_2 \cup I_3) \cap \mathcal{A}(I_2)'$, where the commutant is now interpreted with respect to the action on $H_0(D)$.

By Lemma B.26,

$$D(J_1) \vee \hat{\mathcal{A}}(J_2 \cup I_3) = D(J_1) \vee \big(\mathcal{A}(J_2 \cup I_2 \cup I_3) \cap \mathcal{A}(I_2)'\big)$$

is equal to

$$\big(D(J_1) \vee \mathcal{A}(J_2 \cup I_2 \cup I_3)\big) \cap \mathcal{A}(I_2)' = D(I_1 \cup I_2 \cup I_3) \cap \mathcal{A}(I_2)'$$
$$= \big(D(I_1 \cup I_2 \cup I_3)' \vee \mathcal{A}(I_2)\big)'.$$

The last algebra is equal to $\big(D(I_4) \vee \mathcal{A}(I_2)\big)' = D(I_2 \cup I_4)'$ by Haag duality for defects (Proposition 1.16). □

LEMMA 3.14. *Let I_1, I_2, I_3, I_4 be arranged as in* (3.13). *Assuming D is irreducible, then $\mathcal{A}(I_2)$ is the relative commutant of $\hat{D}(I_1 \cup I_3)$ inside $D(I_1 \cup I_2 \cup I_3)$.*

PROOF. By Lemma B.28, we have $\mathcal{A}(I_2) = \big(\mathcal{A}(I_2) \vee D(I_1 \cup I_2 \cup I_3)'\big) \cap D(I_1 \cup I_2 \cup I_3)$. This algebra is equal to $(\mathcal{A}(I_2) \vee D(I_4)) \cap D(I_1 \cup I_2 \cup I_3) = \hat{D}(I_1 \cup I_3)' \cap D(I_1 \cup I_2 \cup I_3)$. □

In the next Lemma we will use the notion of minimal index $[A : B]$ of a subfactor $B \subseteq A$; see Appendix B.VII for a definition.

LEMMA 3.15. *Let $\mathcal{A}$ be a conformal net with finite index $\mu(\mathcal{A})$, and let ${}_{\mathcal{A}}D_{\mathcal{B}}$ be an irreducible defect. Let I_1, I_2, I_3, I_4 be arranged as in* (3.13). *Then $[\hat{D}(I_1 \cup I_3) : D(I_1 \cup I_3)] \le \mu(\mathcal{A})$.*

PROOF. Note that $\hat{D}(I_1 \cup I_3)$ and $D(I_1 \cup I_3) = D(I_1) \bar{\otimes} \mathcal{A}(I_3)$ are both factors. Let us decompose I_1 into intervals J_1, J_2 as in (3.13). By Lemma 3.12, we have

$$\hat{\mathcal{A}}(J_2 \cup I_3) \vee D(J_1) = \hat{D}(I_1 \cup I_3).$$

We also have

$$\mathcal{A}(J_2 \cup I_3) \vee D(J_1) = D(I_1 \cup I_3).$$

By definition $\mu(\mathcal{A}) = [\hat{\mathcal{A}}(J_2 \cup I_3) : \mathcal{A}(J_2 \cup I_3)]$. The result follows because the minimal index cannot increase under the operation $- \vee D(J_1)$, see (B.20) in Appendix B.VII. □

REMARK 3.16. We will see later, in Corollary 5.20, that in fact we have an equality $[\hat{D}(I_1 \cup I_3) : D(I_1 \cup I_3)] = \mu(\mathcal{A})$.

Finiteness implies semisimplicity. We can now prove the semisimplicity of the fusion of semisimple defects.

PROOF OF THEOREM 3.6. Because the defects D and E are semisimple, we may write them as finite direct sums of irreducible defects: $D = \bigoplus D_i$ and $E = \bigoplus E_j$. Fusion of defects is compatible with direct sums

$$\big(\bigoplus D_i\big) \circledast_{\mathcal{B}} \big(\bigoplus E_j\big) = \bigoplus_{ij} D_i \circledast_{\mathcal{B}} E_j.$$

It therefore suffices to assume D and E are irreducible, and to show that for I genuinely bicolored, the von Neumann algebra $(D \circledast E)(I)$ has finite-dimensional center.

Without loss of generality, we assume that $I = S^1_\top$. Let $H := H_0(D)$ and $K := H_0(E)$ be the vacuum sectors of D and E. The algebra $\tilde{D}(\partial^{\sqsubset}[0,1]^2)$ acts on H (see notation 3.9). Similarly, the algebra $\tilde{E}(\partial^{\sqsupset}[0,1]^2)$ acts on K. Let us denote those two algebras graphically by $\tilde{D}(\sqsubset)$ and $\tilde{E}(\sqsupset)$.

The Hilbert space $H \boxtimes_{\mathcal{B}} K$ is a faithful $(D \circledast E)(S^1_\top)$ – $(D \circledast E)(S^1_\bot)$-bimodule. So by Lemma 3.8, it is enough to show that the algebra of bimodule endomorphisms of $H \boxtimes K$ is finite-dimensional. This algebra of endomorphisms is equal to the algebra of $\tilde{D}(\sqsubset)$-$\tilde{E}(\sqsupset)$-endomorphisms of $H \boxtimes K$. Note that the algebras $\tilde{D}(\sqsubset)$ and $\tilde{E}(\sqsupset)$ are factors by Lemma 3.10. If a bimodule has finite statistical dimension (see Appendix B.VII), then its algebra of bimodule endomorphisms is finite dimensional [**BDH11**, Lem. 4.10]. It is therefore enough to show that the statistical dimension of $H \boxtimes_{\mathcal{B}} K$ as a $\tilde{D}(\sqsubset)$-$\tilde{E}(\sqsupset)$-bimodule is finite.

Using the multiplicativity of the statistical dimension with respect to Connes fusion (B.14) the dimension in question can be computed as

$$\dim\left({}_{\tilde{D}(\sqsubset)}H \boxtimes_{\mathcal{B}} K_{\tilde{E}(\sqsupset)}\right) = \dim\left({}_{\tilde{D}(\sqsubset)}H_{\mathcal{B}(|)}\right) \cdot \dim\left({}_{\mathcal{B}(|)}K_{\tilde{E}(\sqsupset)}\right).$$

So it suffices to argue that the dimension of H as a $\tilde{D}(\sqsubset)$-$\mathcal{B}(|)$-bimodule and the dimension of K as a $\mathcal{B}(|)$-$\tilde{E}(\sqsupset)$-bimodule are finite. This is the content of Lemma 3.17 below. □

Before proceeding, let us fix new names for certain subintervals of our standard circle:

$$I_1 := \partial^{\llcorner}([0, \tfrac{3}{4}] \times [0, \tfrac{1}{2}]) \qquad I_2 := [\tfrac{3}{4}, 1] \times \{0\}$$
$$I_3 := \{1\} \times [0, 1] \qquad I_4 := \partial^{\ulcorner}([0, 1] \times [\tfrac{1}{2}, 1])$$

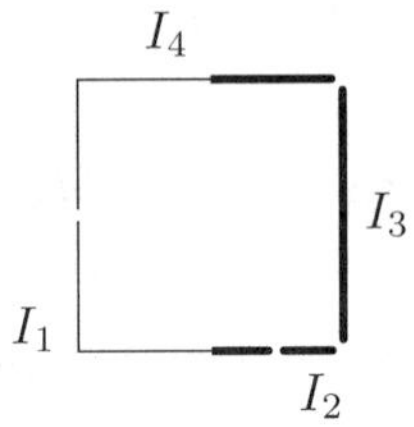

Given a defect ${}_{\mathcal{A}}D_{\mathcal{B}}$, let us also introduce the following shorthand notations:

$$D_{234} := D(I_2 \cup I_3 \cup I_4) \qquad D_{24} := \mathcal{B}(I_2) \mathbin{\bar{\otimes}} D(I_4) \qquad \mathcal{B}_3 := \mathcal{B}(I_3)$$
$$\hat{D}_{24} := \big(D(I_1) \mathbin{\bar{\otimes}} \mathcal{B}(I_3)\big)' \qquad \tilde{D}_{412} := D(I_4) \vee D(I_1 \cup I_2).$$

LEMMA 3.17. *Let $\mathcal{B}$ be a conformal net with finite index and let ${}_{\mathcal{A}}D_{\mathcal{B}}$ be an irreducible defect. Then $H_0(D)$ has finite statistical dimension as a $\tilde{D}_{412}$–$\mathcal{B}_3$-bimodule.*

PROOF. The statistical dimension of $H_0(D)$ as a $\tilde{D}_{412}$–$\mathcal{B}_3$-bimodule is the square root of the minimal index of the subfactor $\mathcal{B}_3 \subseteq \tilde{D}'_{412}$, see (B.15) and the definition of minimal index in Appendix B.VII. Therefore, we need to show that $[\tilde{D}'_{412} : \mathcal{B}_3] < \infty$. We know by Lemma 3.14 (with $\mathcal{A}$ and $\mathcal{B}$ interchanged) that the algebra $\mathcal{B}_3$ is the relative commutant of $\hat{D}_{24}$ inside D_{234}. The algebra $\tilde{D}'_{412}$ is the relative commutant of D_{24} inside D_{234}, as can be seen by taking the commutant of the equation $(D'_{24} \cap D_{234})' = D_{24} \vee D'_{234} = \tilde{D}_{412}$. The index is unchanged by taking commutants, and can only decrease under the operation $- \cap D_{234}$ by (B.19). Thus

$$[\tilde{D}'_{412} : \mathcal{B}_3] = [D'_{24} \cap D_{234} : \hat{D}'_{24} \cap D_{234}] \leq [\hat{D}_{24} : D_{24}],$$

and we have already seen in Lemma 3.15 that $[\hat{D}_{24} : D_{24}] \leq \mu(\mathcal{B}) < \infty$. □

Finiteness of the defect vacuum as a 4-interval bimodule (*splitting* ). We record the following finiteness result, somewhat similar to Lemma 3.15, for future reference.

Let I_1, I_2, I_3, I_4 now be the four sides of our standard bicolored circle:

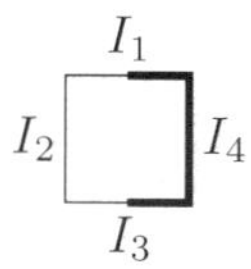

The intervals I_1 and I_3 are genuinely bicolored, I_2 is white, and I_4 is black.

PROPOSITION 3.18. *Let $\mathcal{A}$ and $\mathcal{B}$ be conformal nets with finite index, and let ${}_{\mathcal{A}}D_{\mathcal{B}}$ be an irreducible defect. Then the vacuum sector $H_0(D)$ has finite statistical dimension as a $D(I_1) \vee D(I_3)$ – $(\mathcal{A}(I_2) \vee \mathcal{B}(I_4))^{\mathrm{op}}$-bimodule.*

PROOF. Consider the following intervals:

$J_1 := [\frac{3}{4}, 1] \times \{0\}$ $\quad J_2 := I_4$ $\quad J_3 := [\frac{3}{4}, 1] \times \{1\}$ $\quad J_4 := [\frac{1}{4}, \frac{3}{4}] \times \{1\}$

$J_5 := [0, \frac{1}{4}] \times \{1\}$ $\quad J_6 := I_2$ $\quad J_7 := [0, \frac{1}{4}] \times \{0\}$ $\quad J_8 := [\frac{1}{4}, \frac{3}{4}] \times \{0\}$

which we draw here:

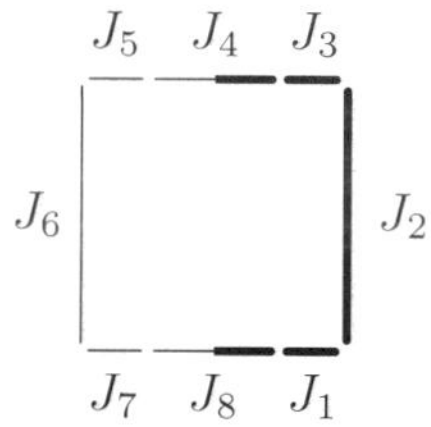

It will be convenient to introduce a graphical notation for the subalgebras of $\mathbf{B}(H_0(D))$ used in this proof:

 $:= D(J_1 \cup J_2 \cup J_3 \cup J_4 \cup J_5 \cup J_6 \cup J_7)$ $:= D(J_4)$

 $:= \mathcal{B}(J_1 \cup J_2 \cup J_3) \vee \mathcal{A}(J_5 \cup J_6 \cup J_7)$ $:= D(J_8)$

 $:= \mathcal{B}(J_1) \vee D(J_3 \cup J_4 \cup J_5) \vee \mathcal{A}(J_7)$  $:= \mathcal{B}(J_2) \vee \mathcal{A}(J_6)$

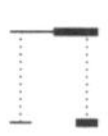 $:= \hat{\mathcal{B}}(J_1 \cup J_3) \vee D(J_4) \vee \hat{\mathcal{A}}(J_5 \cup J_7)$  $:= \mathcal{B}(J_2) \vee \mathcal{A}(J_6) \vee D(J_8)$

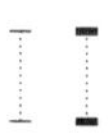 $:= \hat{\mathcal{B}}(J_1 \cup J_3) \vee \hat{\mathcal{A}}(J_5 \cup J_7)$  $:= \mathcal{B}(J_1 \cup J_3) \vee \mathcal{A}(J_5 \cup J_7)$,

where, as in 3.11, $\hat{\mathcal{A}}(J_5 \cup J_7)$ is the relative commutant of $\mathcal{A}(J_6)$ in $\mathcal{A}(J_5 \cup J_6 \cup J_7)$ and $\hat{\mathcal{B}}(J_1 \cup J_3)$ is the relative commutant of $\mathcal{B}(J_2)$ in $\mathcal{B}(J_1 \cup J_2 \cup J_3)$. Note that  is the commutant of  since, by Lemma B.26, we have

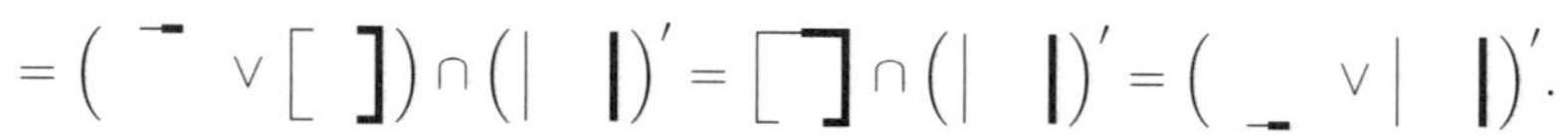

In particular, the algebra  is a factor.

We have to show that

$$\left[(D(I_1) \vee D(I_3))' : \mathcal{A}(I_2) \vee \mathcal{B}(I_4)\right] < \infty.$$

Using Haag duality and strong additivity, note that the algebra $(D(I_1) \vee D(I_3))'$ is the relative commutant of  inside . Similarly, it follows from Lemma B.28 that the algebra $\mathcal{A}(I_2) \vee \mathcal{B}(I_4)$ is the relative commutant of  in :

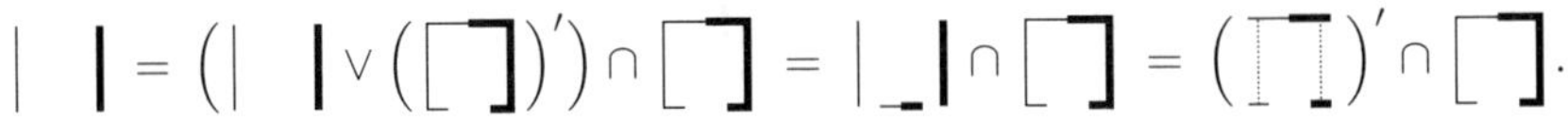

By (B.19) and (B.20), we then have

$$\left[(D(I_1) \vee D(I_3))' : \mathcal{A}(I_2) \vee \mathcal{B}(I_4)\right] =$$

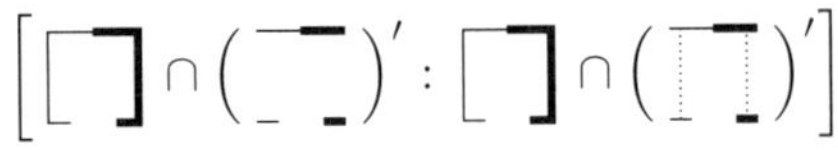

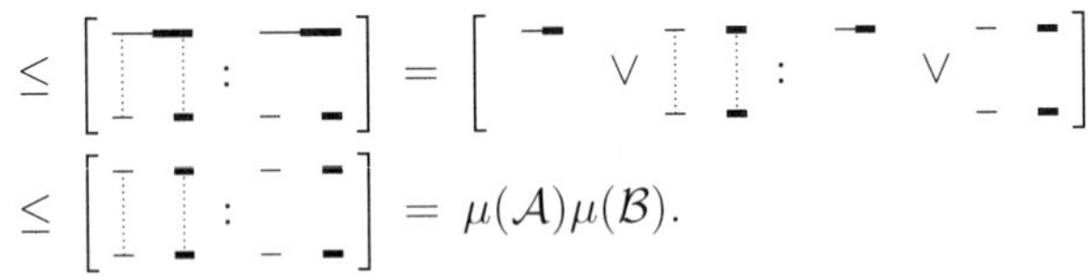

$$= \mu(\mathcal{A})\mu(\mathcal{B}).$$

□

CHAPTER 4

A variant of horizontal fusion

In Section 2.B we saw how to define the horizontal fusion of two sectors. We will now define a variant of the horizontal fusion, called keystone fusion, which itself depends on an intermediate construction we refer to as keyhole fusion. In Section 4.D, we will show that horizontal fusion and keystone fusion are in fact naturally isomorphic, and we will construct a canonical isomorphism Φ between them. That isomorphism will be essential in our construction of the $1\boxtimes 1$-isomorphism Ω (4.31).

Recall that we implicitly assume that all our conformal nets are irreducible.

4.A. The keyhole and keystone fusion

Recall the Notation 1.49. Consider the intervals $I_l := \partial^{\sqsupset}([2/3,1]\times[0,1])$, $I_r := \partial^{\sqsubset}([1,4/3]\times[0,1])$, and $I := I_l \cap I_r$. Orient I_l and I_r counterclockwise, and orient I so that the inclusion $I \hookrightarrow I_r$ is orientation preserving—see (4.1). The inclusion $I \hookrightarrow I_l$ is then orientation reversing. Let J be the closure of $(I_l \cup I_r)\setminus I$. We orient J so that it agrees with the orientation of I_l on $J\cap I_l$. We draw these intervals as follows:

(4.1) $I_l =$ , $I_r =$ , $I =$  and $J =$ .

Given a conformal net $\mathcal{A}$ with finite index, we will define three functors

$$F,\ G_0,\ G\ :\ \mathcal{A}(I_l)\text{-modules} \times \mathcal{A}(I_r)\text{-modules} \ \to\ \mathcal{A}(J)\text{-modules}.$$

These operations will be called respectively the fusion, the keyhole fusion, and the keystone fusion, and will be denoted graphically as follows:

$F(H_l,H_r) =$ 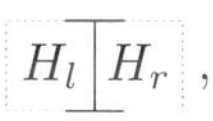

, $G_0(H_l,H_r) =$ 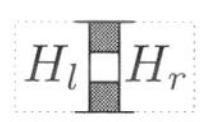

, $G(H_l,H_r) =$ 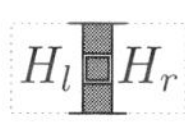

.

When we want to stress the dependence on the conformal net $\mathcal{A}$, we will denote these functors $F_{\mathcal{A}}$, $G_{0,\mathcal{A}}$, $G_{\mathcal{A}}$.

The ordinary horizontal fusion. The functor F is defined by fusion over $\mathcal{A}(I)$: using the orientation preserving inclusion $I \hookrightarrow I_r$, any left $\mathcal{A}(I_r)$-module is also a left $\mathcal{A}(I)$-module, and using the orientation reversing inclusion $I_0 \hookrightarrow I_l$, any left $\mathcal{A}(I_l)$-module is also a right $\mathcal{A}(I)$-module. We can therefore define the horizontal fusion functor as follows:

$$F(H_l,H_r) := H_l \boxtimes_{\mathcal{A}(I)} H_r.$$

Write J as $J_1 \sqcup J_2$; we obtain actions of $\mathcal{A}(J_1)$ and $\mathcal{A}(J_2)$ on $H_l \boxtimes_{\mathcal{A}(I)} H_r$, by [**BDH13**, Cor. 1.29]. Note that in the case $H_l = L^2\mathcal{A}(I_l)$ and $H_r = L^2\mathcal{A}(I_r)$, the actions of $\mathcal{A}(J_1)$ and $\mathcal{A}(J_2)$ extend to an action of $\mathcal{A}(J) = \mathcal{A}(J_1)\,\bar{\otimes}\,\mathcal{A}(J_2)$; the same therefore holds for arbitrary H_l and H_r. The only difference between the functor F and the functor $\mathsf{fusion_h}$ from (2.12) is that they have somewhat different source and target categories—the main construction is identical in both functors.

The keyhole fusion. We will need to name a few more manifolds. Let

$$\tilde{I}_l := \partial^{\sqsupset}([2/3,5/6]\times[0,1]) \qquad \tilde{I}_r := \partial^{\sqsubset}([7/6,4/3]\times[0,1])$$
$$S_u := \partial([5/6,7/6]\times[2/3,1]) \qquad S_m := \partial([5/6,7/6]\times[1/3,2/3])$$
$$S_d := \partial([5/6,7/6]\times[0,1/3]) \qquad K := (S_u \cup S_d)\cap(\tilde{I}_l\cup\tilde{I}_r).$$

We draw these as

(4.2) $\tilde{I}_l =$ 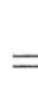 , $\tilde{I}_r =$ , $S_u =$ , $S_m =$ , $S_d =$  and $K =$ 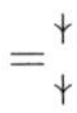.

The intervals $\tilde{I}_l$ and $\tilde{I}_r$ are oriented counterclockwise, as were I_l and I_r. The manifolds S_u, S_m and S_d are conformal circles via their constant speed parametrizations and are also oriented counterclockwise. (A conformal circle is a circle together with a homeomorphism with S^1 that is only determined up to orientation preserving conformal diffeomorphisms of S^1.) Finally, the manifold K inherits its orientation from $S_u \cup S_d$. Note that the inclusion $K \hookrightarrow \tilde{I}_l \cup \tilde{I}_r$ is orientation reversing. We will also need the reflection j along the horizontal axis $y = 1/2$.

Let us fix orientation preserving identifications $\phi_l : \tilde{I}_l \xrightarrow{\cong} I_l$ and $\phi_r : \tilde{I}_r \xrightarrow{\cong} I_r$ that are symmetric with respect to the reflection j, restrict to the identity in a neighborhood of $\partial\tilde{I}_l = \partial I_l$ and $\partial\tilde{I}_r = \partial I_r$, and satisfy $\phi_l(5/6,t) = (1,t)$ and $\phi_r(7/6,t) = (1,t)$ for all $t \in [0,1]$. Using these identifications, any $\mathcal{A}(I_l)$-module becomes an $\mathcal{A}(\tilde{I}_l)$-module and any $\mathcal{A}(I_r)$-module becomes an $\mathcal{A}(\tilde{I}_r)$-module. We can now define the keyhole fusion functor as follows:

$$G_0(H_l,H_r) := \big(H_l \otimes H_r\big) \underset{\mathcal{A}(K)}{\boxtimes} \big(H_0(S_u)\otimes H_0(S_d)\big),$$

where $H_0(S_u)$ and $H_0(S_d)$ are the canonical vacuum sectors; see (C.1). The right-hand side is an instance of what we call cyclic fusion—see Appendix B.III. In the notation of cyclic fusion, we have

$$G_0(H_l,H_r) = H_l \underset{\mathcal{A}(K_1)}{\boxtimes} H_0(S_u) \underset{\mathcal{A}(K_2)}{\boxtimes} H_r \underset{\mathcal{A}(K_3)}{\boxtimes} H_0(S_d) \underset{\mathcal{A}(K_4)}{\boxtimes}$$

where $K_1 = \tilde{I}_l \cap S_u$, $K_2 = \tilde{I}_r \cap S_u$, $K_3 = \tilde{I}_r \cap S_d$, and $K_4 = \tilde{I}_l \cap S_d$, appropriately oriented. It follows from [**BDH13**, Cor. 1.29] that the algebras $\mathcal{A}(J\cap(\tilde{I}_l\cup\tilde{I}_r))$ and $\mathcal{A}(J\cap(S_u\cup S_d))$ generate an action of $\mathcal{A}(J)$ on $G_0(H_l,H_r)$.

The keystone fusion. Note that the algebra

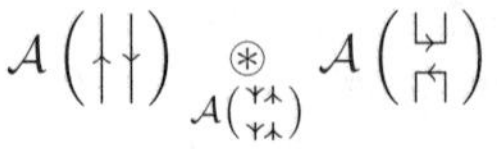

has a natural left action on $G_0(H_l,H_r)$ commuting with the action of $\mathcal{A}(J)$. This algebra can be identified with $\mathcal{A}(S_m)^{op} \cong \mathcal{A}(-S_m)$, where $-S_m$ denotes the circle S_m equipped with the opposite (i.e., clockwise) orientation. Here, we use the extension of $\mathcal{A}$ from intervals to 1-manifolds constructed in [**BDH14**, Sec. 1.A]; see also Appendix C.VI. By definition,[10] the algebra $\mathcal{A}(S_m)$ is generated by all the $\mathcal{A}(I)$, for $I \subset S_m$, acting on the Hilbert space .

[10] For this, we implicitly identify the surfaces 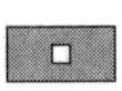 and $S_m \times [0,1]$.

By Theorem C.9, the algebra $\mathcal{A}(S_m)$ contains a direct summand that is canonically isomorphic to $\mathbf{B}(H_0(S_m, \mathcal{A}))$. We can therefore define the keystone fusion functor as follows:

$$G(H_l, H_r) := G_0(H_l, H_r) \boxtimes_{\mathcal{A}(S_m)} H_0(S_m).$$

Moreover, since $\mathbf{B}(H_0(S_m, \mathcal{A}))$ and $\mathcal{A}(J)$ commute on $G_0(H_l, H_r)$, there is a residual action of $\mathcal{A}(J)$ on the Hilbert space $G(H_l, H_r)$.

Fusion and keystone fusion are isomorphic. We will show presently that the functors

$$F, G : \mathcal{A}(I_l)\text{-modules} \times \mathcal{A}(I_r)\text{-modules} \to \mathcal{A}(J)\text{-modules}$$

are naturally isomorphic to one another, and then later (in Proposition 4.29) construct a specific such natural isomorphism.

We use the following straightforward generalization of Lemma B.24.

LEMMA 4.3. *Let* $F, G : A_1\text{-modules} \times A_2\text{-modules} \to B\text{-modules}$ *be normal functors (see Appendix* B.VIII*), and let* M_i *be a faithful* A_i*-module, for* $i = 1, 2$*. Then, in order to uniquely define a natural transformation* $a : F \to G$*, it is enough to specify its value on* (M_1, M_2) *and to check that for each pair* (r_1, r_2) *with* $r_i \in \mathrm{End}_A(M_i)$*, the diagram*

$$\begin{array}{ccc} F(M_1, M_2) & \xrightarrow{F(r_1, r_2)} & F(M_1, M_2) \\ {\scriptstyle a_{M_1,M_2}}\downarrow & & \downarrow{\scriptstyle a_{M_1,M_2}} \\ G(M_1, M_2) & \xrightarrow{G(r_1, r_2)} & G(M_1, M_2) \end{array}$$

commutes. □

Using Theorem C.8 and the above lemma, we prove that the two different versions of horizontal fusion are naturally isomorphic to each other:

PROPOSITION 4.4. *There exists a natural isomorphism between the fusion functor* F *and the keystone fusion functor* G*.*

PROOF. Consider the circles

$$S_l := \partial([0,1]\times[0,1]), \qquad S_r := \partial([1,2]\times[0,1]) \qquad S_b := \partial([0,2]\times[1,2])$$
$$\tilde{S}_l := \partial([0,5/6] \times [0,1]), \quad \tilde{S}_r := \partial([7/6,2] \times [0,1])$$
$$\hat{S}_l := \partial([0,5/6] \times [0,1] \cup [5/6,7/6] \times [0,1/3] \cup [5/6,7/6] \times [2/3,1])$$

which we draw as follows:

(4.5)

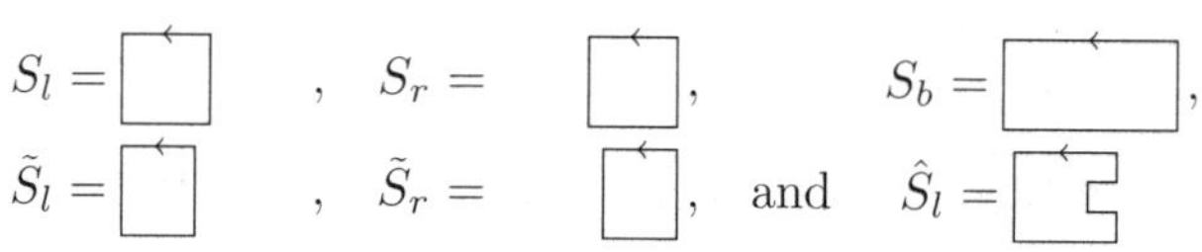

('b' stands for big). The identifications $\phi_l : \tilde{I}_l \xrightarrow{\cong} I_l$ and $\phi_r : \tilde{I}_r \xrightarrow{\cong} I_r$ induce isomorphisms $H_0(\tilde{S}_l) \cong H_0(S_l)$ and $H_0(\tilde{S}_r) \cong H_0(S_r)$ that are equivariant with respect to $\mathcal{A}(I_l')$ and $\mathcal{A}(I_r')$ (here, I_l' and I_r' are the closures of $S_l \setminus I_l$ and $S_r \setminus I_r$, respectively). From the isomorphism $H_0(\hat{S}_l) \cong H_0(S_d) \boxtimes_{\mathcal{A}(K_4)} H_0(\tilde{S}_l) \boxtimes_{\mathcal{A}(K_1)} H_0(S_u)$

it follows that $G_0\big(H_0(S_l), H_0(S_r)\big)$ represents the Hilbert space of an annulus; see Appendix C.V. Using Theorem C.8 we therefore have

$$\begin{aligned} G_0\big(H_0(S_l), H_0(S_r)\big) &= \Big(H_0(\tilde{S}_l) \underset{\mathcal{A}(K_1)}{\boxtimes} H_0(S_u) \underset{\mathcal{A}(K_2)}{\boxtimes} H_0(\tilde{S}_r) \underset{\mathcal{A}(K_3)}{\boxtimes} H_0(S_d) \underset{\mathcal{A}(K_4)}{\boxtimes} \Big) \\ &\cong \Big(H_0(\hat{S}_l) \underset{\mathcal{A}(K_2)}{\boxtimes} H_0(\tilde{S}_r) \underset{\mathcal{A}(K_3)}{\boxtimes} \Big) \\ &\cong \bigoplus_{\lambda \in \Delta} H_\lambda(-S_m) \otimes H_{\bar{\lambda}}(S_b). \end{aligned}$$

We draw the above isomorphisms as follows:

$$G_0\big(H_0(S_l), H_0(S_r)\big) =$$

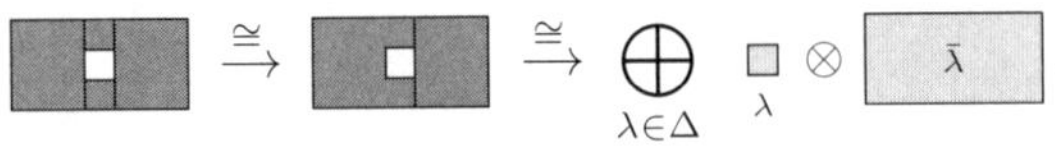

Note that the two isomorphisms intertwine the natural actions of $\{\mathcal{A}(I)\}_{I \subset S_b}$ and $\{\mathcal{A}(I)\}_{I \subset -S_m}$. We can now compute

$$\begin{aligned} G\big(H_0(S_l), H_0(S_r)\big) &:= G_0\big(H_0(S_l), H_0(S_r)\big) \boxtimes_{\mathcal{A}(S_m)} H_0(S_m) \\ &\cong \Big(\bigoplus_{\lambda \in \Delta} H_\lambda(-S_m) \otimes H_{\bar{\lambda}}(S_b) \Big) \boxtimes_{\mathcal{A}(S_m)} H_0(S_m) \\ &\cong \Big(H_0(-S_m) \otimes H_0(S_b) \Big) \boxtimes_{\mathbf{B}(H_0(S_m))} H_0(S_m) \qquad (4.6) \\ &\cong H_0(S_b) \otimes \Big(H_0(-S_m) \boxtimes_{\mathbf{B}(H_0(S_m))} H_0(S_m) \Big) \\ &\cong H_0(S_b) \otimes \mathbb{C} \;\cong\; H_0(S_b). \end{aligned}$$

Combining (4.6) with the non-canonical isomorphism $F(H_0(S_l), H_0(S_r)) \cong H_0(S_b)$ from (C.5), we get an isomorphism

$$\varphi : G(H_0(S_l), H_0(S_r)) \xrightarrow{\cong} F(H_0(S_l), H_0(S_r)),$$

compatible with the actions of $\mathcal{A}(I_l')$ and of $\mathcal{A}(I_r')$.

Since $H_0(S_l)$ and $H_0(S_r)$ are faithful $\mathcal{A}(I_l)$- and $\mathcal{A}(I_r)$-modules, we can use Lemma 4.3 to finish the argument: it remains only to check that φ is equivariant with respect to all $r_1 \in \mathrm{End}_{\mathcal{A}(I_l)}(H_0(S_l))$ and $r_2 \in \mathrm{End}_{\mathcal{A}(I_r)}(H_0(S_r))$. That equivariance follows immediately from Haag duality for nets (Proposition C.4) and the fact that φ is equivariant with respect to $\mathcal{A}(I_l')$ and $\mathcal{A}(I_r')$. □

Unfortunately, the above proposition is not sufficient for our purposes: it does not construct a natural isomorphism $\Phi_{\mathcal{A}} : F_{\mathcal{A}} \to G_{\mathcal{A}}$, but only proves that one exists. This leaves unsettled, for instance, the question of whether these natural isomorphisms can be chosen so that $\Phi_{\mathcal{A}\otimes\mathcal{B}} = \Phi_{\mathcal{A}} \otimes \Phi_{\mathcal{B}}$. In the following sections, we will construct a canonical choice of such natural isomorphisms for which the desired symmetric monoidal property is clear.

4.B. The keyhole fusion of vacuum sectors of defects

Let S_l, S_r, S_b, $\tilde{S}_l$, $\tilde{S}_r$, S_u, S_m, S_d, I_l, I_r, $\tilde{I}_l$, $\tilde{I}_r$, K be as in (4.1), (4.2), and (4.5). We bicolor S_l, S_r, $\tilde{S}_l$, $\tilde{S}_r$ by setting

$$\begin{array}{llll} (S_l)_\circ := (S_l)_{x \le \frac{1}{2}} & (S_l)_\bullet := (S_l)_{x \ge \frac{1}{2}} & (S_r)_\circ := (S_r)_{x \le \frac{3}{2}} & (S_r)_\bullet := (S_r)_{x \ge \frac{3}{2}} \\ (\tilde{S}_l)_\circ := (\tilde{S}_l)_{x \le \frac{1}{2}} & (\tilde{S}_l)_\bullet := (\tilde{S}_l)_{x \ge \frac{1}{2}} & (\tilde{S}_r)_\circ := (\tilde{S}_r)_{x \le \frac{3}{2}} & (\tilde{S}_r)_\bullet := (\tilde{S}_r)_{x \ge \frac{3}{2}}. \end{array} \qquad (4.7)$$

Denote by j the reflection across the horizontal axis $y = 1/2$, and let

$$S_{l,\top} := (S_l)_{y\geq\frac{1}{2}} \qquad S_{r,\top} := (S_r)_{y\geq\frac{1}{2}} \qquad S_{b,\top} := (S_b)_{y\geq\frac{1}{2}}$$
$$\tilde{S}_{l,\top} := (\tilde{S}_l)_{y\geq\frac{1}{2}} \quad \tilde{S}_{r,\top} := (\tilde{S}_r)_{y\geq\frac{1}{2}} \quad \tilde{I}_{l,\top} := (\tilde{I}_l)_{y\geq\frac{1}{2}} \quad \tilde{I}_{r,\top} := (\tilde{I}_r)_{y\geq\frac{1}{2}}.$$

Let $\mathcal{A}$, $\mathcal{B}$, $\mathcal{C}$ be conformal nets, and let ${}_{\mathcal{A}}D_{\mathcal{B}}$ and ${}_{\mathcal{B}}E_{\mathcal{C}}$ be defects. We are interested in evaluating $G_0 := G_{0,\mathcal{B}}$ on the vacuum sectors

$$H_0(S_l, D) := L^2(D(S_{l,\top})) \qquad \text{and} \qquad H_0(S_r, E) := L^2(E(S_{r,\top}))$$

from Definition 1.50. These have compatible actions of the algebras $\{D(I)\}_{I\subset S_l}$ and $\{E(I)\}_{I\subset S_r}$ respectively. In particular, they are respectively $\mathcal{B}(I_l)$- and $\mathcal{B}(I_r)$-modules, and so we can apply the functor G_0.

Let us also define

$$H_0(\tilde{S}_l, D) := L^2(D(\tilde{S}_{l,\top})) \qquad \text{and} \qquad H_0(\tilde{S}_r, E) := L^2(E(\tilde{S}_{r,\top})).$$

Recall that the definition of the functor G_0 uses identifications $\phi_l : \tilde{I}_l \xrightarrow{\cong} I_l$ and $\phi_r : \tilde{I}_r \xrightarrow{\cong} I_r$ to endow $H_0(S_l, D)$ with a $\mathcal{B}(\tilde{I}_l)$-action, and $H_0(S_r, D)$ with a $\mathcal{B}(\tilde{I}_r)$-action. We write $\phi_l^* H_0(S_l, D)$ and $\phi_r^* H_0(S_l, D)$ for the resulting $\tilde{S}_l$-sector of D and $\tilde{S}_r$-sector of E. (Here, given a bicolored circle S and a defect D, a Hilbert space is called an S-sector of D if it has compatible actions of $D(I)$ for all bicolored intervals $I \subset S$—compare Appendix C.V.) Recall that the maps ϕ_l and ϕ_r were chosen to commute with j, and to be the identity near to the boundary. Let us call

$$\phi_{l,\top} : \tilde{S}_{l,\top} \xrightarrow{\cong} S_{l,\top}\,, \qquad \phi_{r,\top} : \tilde{S}_{r,\top} \xrightarrow{\cong} S_{r,\top}$$

the extension by the identity of the maps $\phi_l|_{\tilde{I}_{l,\top}}$ and $\phi_r|_{\tilde{I}_{r,\top}}$. We then have canonical identifications

$$L^2(D(\phi_{l,\top})) \;:\; H_0(\tilde{S}_l, D) \to \phi_l^* H_0(S_l, D)$$
$$L^2(D(\phi_{r,\top})) \;:\; H_0(\tilde{S}_r, D) \to \phi_r^* H_0(S_r, D)$$

of $\tilde{S}_l$-sectors of D and $\tilde{S}_r$-sectors of E.

We now have an isomorphism

$$\begin{aligned} &G_{0,\mathcal{B}}\big(H_0(S_l, D), H_0(S_r, E)\big) \\ &\quad\cong \big(H_0(\tilde{S}_l, D)\otimes H_0(\tilde{S}_r, E)\big) \underset{\mathcal{B}(K)}{\boxtimes} \big(H_0(S_u, \mathcal{B})\otimes H_0(S_d, \mathcal{B})\big) \\ &\quad= H_0(\tilde{S}_l, D) \underset{\mathcal{B}(K_1)}{\boxtimes} H_0(S_u, \mathcal{B}) \underset{\mathcal{B}(K_2)}{\boxtimes} H_0(\tilde{S}_r, E) \underset{\mathcal{B}(K_3)}{\boxtimes} H_0(S_d, \mathcal{B}) \underset{\mathcal{B}(K_4)}{\boxtimes} \end{aligned}$$

which we draw as follows:

(4.8) $$G_{0,\mathcal{B}}\big(H_0(S_l, D), H_0(S_r, E)\big) \cong$$ 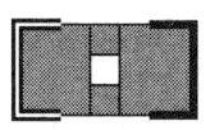 .

Here, the lines ═, —, and ▬ correspond to the conformal nets $\mathcal{A}$, $\mathcal{B}$, and $\mathcal{C}$, and the transition points ═—, and —▬ indicate the defects D and E.

Keyhole fusion as an L^2-space. We need to introduce yet more manifolds. We have already encountered $K_1 = \tilde{S}_l \cap S_u$ and $K_2 = \tilde{S}_r \cap S_u$. We define $K_u := K_1 \cup K_2$ and $J_u := J_1 \cup J_2$, where $J_1 := S_b \cap S_u$ and $J_2 := S_u \cap S_m$. We orient K_u and J_u compatibly with S_u. Let J_l be the closure of $\tilde{S}_{l,\top} \setminus K_1$ and, similarly, let J_r be the closure of $\tilde{S}_{r,\top} \setminus K_2$. The orientations and the bicolorings of J_l and J_r are inherited from $\tilde{S}_l$ and $\tilde{S}_r$. We include pictures of these manifolds:

(4.9) $J_l =$  , $J_r =$  , $K_u =$  , $J_u =$.

Following Notation 3.11, we let $\hat{\mathcal{B}}(J_u)$ denote the commutant of $\mathcal{B}(K_u)$ on $H_0(S_u, \mathcal{B})$. Our computation of the keyhole fusion will be in terms of the algebra

$$D(J_l) \vee \hat{\mathcal{B}}(J_u) \vee E(J_r) \subset \mathbf{B}\big(H_0(\tilde{S}_l, D) \underset{\mathcal{B}(K_1)}{\boxtimes} H_0(S_u, \mathcal{B}) \underset{\mathcal{B}(K_2)}{\boxtimes} H_0(\tilde{S}_r, E)\big),$$

which we denote pictorially by

(4.10) $D(J_l) \vee \hat{\mathcal{B}}(J_u) \vee E(J_r) =$  .

The dotted line in this picture picture serves to remind us that $\hat{\mathcal{B}}(J_u)$ was used instead of $\mathcal{B}(J_u)$. Note that the algebra (4.10) also acts on $G_{0,\mathcal{B}}(H_0(S_l, D), H_0(S_r, E))$ because that fusion is obtained from

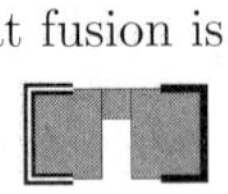 $= H_0(\tilde{S}_l, D) \boxtimes_{\mathcal{B}(K_1)} H_0(S_u, \mathcal{B}) \boxtimes_{\mathcal{B}(K_2)} H_0(\tilde{S}_r, E)$

by fusing it over $\mathcal{B}(K_3 \cup K_4)$ with $H_0(S_d, \mathcal{B})$.

Let

$$\begin{aligned} S_{b,\ulcorner} &:= \{(x,y) \in S_b \,|\, x - y < \tfrac{1}{2}\} \\ S_{b,\llcorner} &:= \{(x,y) \in S_b \,|\, x + y < \tfrac{3}{2}\} \\ S_{b,\urcorner} &:= \{(x,y) \in S_b \,|\, x + y > \tfrac{3}{2}\} \\ S_{b,\lrcorner} &:= \{(x,y) \in S_b \,|\, x - y > \tfrac{1}{2}\}, \end{aligned}$$

with orientations and bicolorings as in the following pictures

$S_{b,\ulcorner} =$ 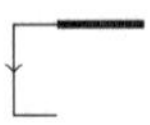 , $S_{b,\llcorner} =$ 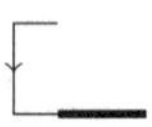 , $S_{b,\urcorner} =$ 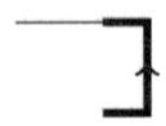 , $S_{b,\lrcorner} =$  .

Note that these manifolds do not include their boundary points.

THEOREM 4.11. *Let $\mathcal{A}$, $\mathcal{B}$, $\mathcal{C}$ be conformal nets, and let ${}_{\mathcal{A}}D_{\mathcal{B}}$ and ${}_{\mathcal{B}}E_{\mathcal{C}}$ be defects. Then there is a canonical unitary isomorphism*

(4.12) $\Psi_0 : L^2\Big($  $\Big) \xrightarrow{\cong}$ 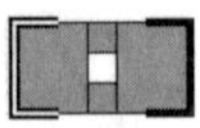 .

In formulas, this is a map

$$\Psi_0 = (\Psi_0)_{D,E} \colon L^2\big(D(J_l) \vee \hat{\mathcal{B}}(J_u) \vee E(J_r)\big) \xrightarrow{\cong} G_{0,\mathcal{B}}\big(H_0(S_l, D), H_0(S_r, E)\big),$$

where S_l, S_r, J_l, J_u, J_r are as in (4.7) and (4.9). The map Ψ_0 is equivariant for the natural left actions of $\{D(I)\}_{I \subset S_{b,\ulcorner}}$, $\{D(I)\}_{I \subset S_{b,\llcorner}}$, $\{E(I)\}_{I \subset S_{b,\urcorner}}$, and $\{E(I)\}_{I \subset S_{b,\lrcorner}}$, and for the natural right actions of $\{\mathcal{B}(I)\}_{I \subset S_m}$.

PROOF. This is the special case of Proposition B.34 where $M := D(\tilde{S}_{l,\top}) \,\bar{\otimes}\, E(\tilde{S}_{r,\top})$, $M_0 = D(J_l) \,\bar{\otimes}\, E(J_r)$, $A = \mathcal{B}(K_u)^{op}$, and $H = H_0(S_u, \mathcal{B})$. In pictures, these are

(4.13) $M :=$  , $M_0 :=$  , $A :=$  , and $H :=$  .

The equivariance of Ψ_0 is clear for intervals I that are contained in the upper half $\{(x,y) | y \geq \frac{1}{2}\}$ or in the lower half $\{(x,y) | y \leq \frac{1}{2}\}$, and follows by strong additivity for more general intervals. □

Associativity of the L^2-space identification. The isomorphism Ψ_0 is in an appropriate sense associative, as follows. Suppose that we have three defects ${}_{\mathcal{A}}D_{\mathcal{B}}$, ${}_{\mathcal{B}}E_{\mathcal{C}}$, and ${}_{\mathcal{C}}F_{\mathcal{D}}$. We then have various applications of Ψ_0 forming the square

(4.14)

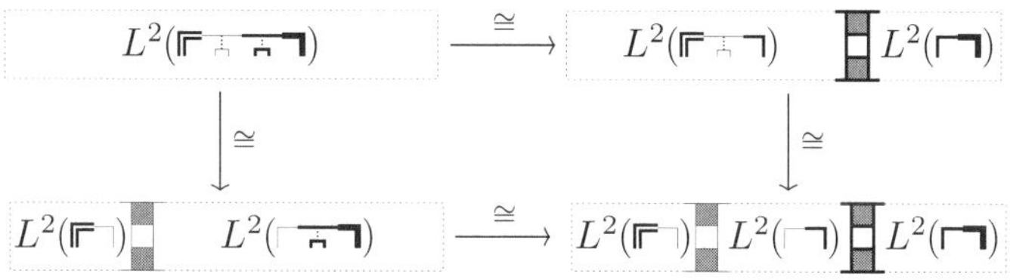

This diagram commutes by Proposition B.37.

We explain the meaning of the pictures in this square. The conformal nets $\mathcal{A}$, $\mathcal{B}$, $\mathcal{C}$, $\mathcal{D}$ are indicated by lines of various thickness, and the defects ${}_{\mathcal{A}}D_{\mathcal{B}}$, ${}_{\mathcal{B}}E_{\mathcal{C}}$, ${}_{\mathcal{C}}F_{\mathcal{D}}$ are indicated by the transitions. The notations $L^2(\ )$, $L^2(\ )$, $L^2(\ )$, stand for $L^2(D(S^1_\top))$, $L^2(E(S^1_\top))$, $L^2(F(S^1_\top))$. The lower right corner of (4.14) is

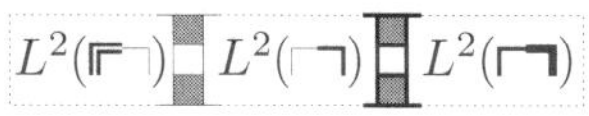

$$= G_{0,\mathcal{B}}\big(L^2(D(S^1_\top)), G_{0,\mathcal{C}}\big(L^2(E(S^1_\top)), L^2(F(S^1_\top))\big)\big)$$
$$= G_{0,\mathcal{C}}\big(G_{0,\mathcal{B}}\big(L^2(D(S^1_\top)), L^2(E(S^1_\top))\big), L^2(F(S^1_\top))\big).$$

Note that, following (4.8), this Hilbert space is also denoted.

As in (4.10), $L^2(\ )$ and $L^2(\ )$ denote the Hilbert spaces $L^2(D(J_l) \vee \hat{\mathcal{B}}(J_u) \vee E(J_r))$ and $L^2(E(J_l) \vee \hat{\mathcal{C}}(J_u) \vee F(J_r))$, respectively. The upper right and lower left corners of (4.14) are therefore given by

$$= G_{0,\mathcal{C}}\big(L^2(D(J_l) \vee \hat{\mathcal{B}}(J_u) \vee E(J_r)), L^2(F(S^1_\top))\big)$$

and

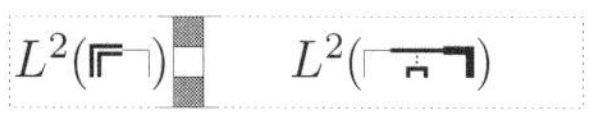

$$= G_{0,\mathcal{B}}\big(L^2(D(S^1_\top)), L^2(E(J_l) \vee \hat{\mathcal{C}}(J_u) \vee F(J_r))\big).$$

Finally, the vector space $L^2(\ )$ that appears in the upper left corner of (4.14) is the L^2 space of the von Neumann algebra

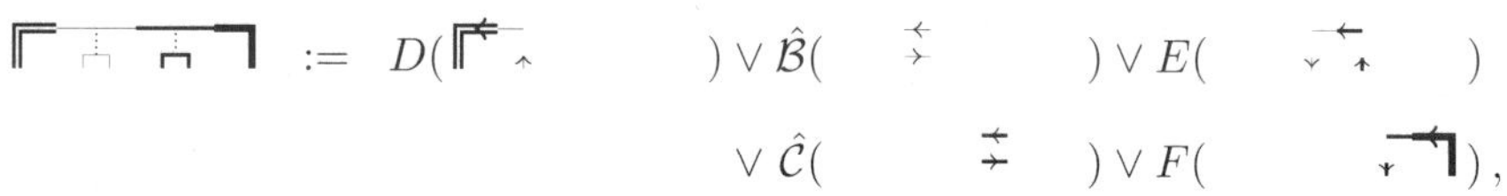

where the completion is taken on the Hilbert space or, equivalently, on the Hilbert space.

4.C. The keystone fusion of vacuum sectors of defects

In this section, the defects ${}_{\mathcal{A}}D_{\mathcal{B}}$ and ${}_{\mathcal{B}}E_{\mathcal{C}}$ are assumed to be irreducible. As before, the conformal net $\mathcal{B}$ is taken to be of finite index.

Recall the algebra

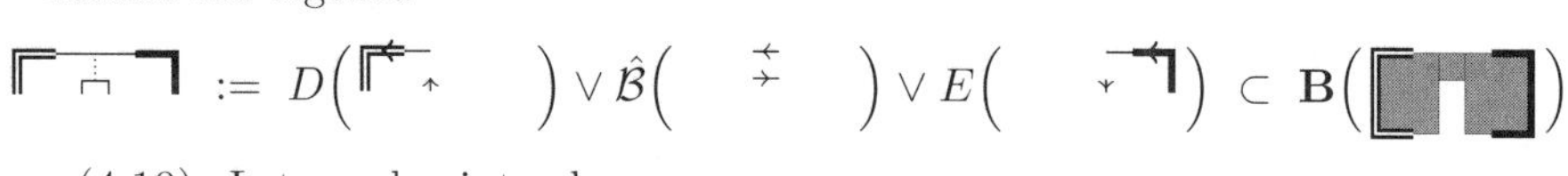

from (4.10). Let us also introduce

The algebra 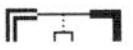 is a factor, as can be seen by applying Lemma B.36 in the situation of (4.13), but its subalgebra  will typically not be a factor. However, since $\mathcal{B}$ has finite index, we know by Theorem 3.6 that the subalgebra has finite dimensional center.

LEMMA 4.15. *Let $\mathcal{B}$ be a conformal net with finite index $\mu(\mathcal{B})$, and let ${}_{\mathcal{A}}D_{\mathcal{B}}$ and ${}_{\mathcal{B}}E_{\mathcal{C}}$ be irreducible defects. Let $p_1, \ldots, p_n$ be the minimal central projections of the algebra* 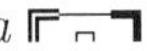. *Then we have*

$$\sum_i \Big[p_i$$ 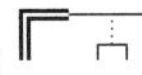 $$p_i \; : \; p_i$$ 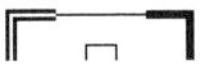 $$\Big] \leq \mu(\mathcal{B}).$$

PROOF. To simplify the notation we abbreviate $N := \mathcal{B}($  $)$, $M := \hat{\mathcal{B}}($  $)$, and $A := D($ 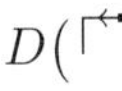 $) \bar{\otimes} E($  $)$. Then  $= M \vee A$ and 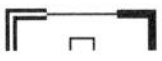 $= N \vee B$. By definition, $\mu(\mathcal{B}) = [M : N]$ is the square of $[\![M : N]\!]$, the latter being our notation for the statistical dimension of ${}_N L^2(M)_M$. Also, $[p_i(M \vee A)p_i : p_i(N \vee A)]$ is the square of the statistical dimension of ${}_{p_i(N\vee A)}p_i L^2(M \vee A)_{M\vee A}$, as can be seen using (B.15); cf. the proof of [**BDH11**, Prop. 5.18]. The vector whose entries are the statistical dimensions of ${}_{p_i(N\vee A)}p_i L^2(M \vee A)_{M\vee A}$ is denoted by $[\![M \vee A : N \vee A]\!]$. By (B.20)

$$\big\| [\![M \vee A : N \vee A]\!] \big\|_2 \;\leq\; [\![M : N]\!]$$

and the result follows. □

COROLLARY 4.16. *The inclusion*  $\hookrightarrow$  *is a finite homomorphism of von Neumann algebras with finite-dimensional center (see Appendix* B.VI*).*

PROOF. Let $X :=$ 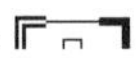, with minimal central projections $p_1, \ldots, p_n$, and let $Y :=$ . Recall that Y is a factor. By definition, the inclusion $X \to Y$ is finite if and only if the bimodule ${}_X L^2 Y_Y$ is dualizable, which happens if and only if its summands ${}_{p_i X}(p_i L^2 Y)_Y$ are dualizable. A bimodule between factors is dualizable if and only if its statistical dimension is finite; see Appendix B.VII. The commutant of Y on $p_i L^2 Y$ is $p_i Y p_i$, and the inclusion $p_i X \hookrightarrow p_i Y p_i$ is finite by the previous lemma. □

Keystone fusion contains the vacuum sector of the fused defect. Let $\mathcal{B}$ be a conformal net with finite index. Recall from Section 4.A that, given a $\mathcal{B}(I_l)$-module H_l and a $\mathcal{B}(I_r)$-module H_r, then the keystone fusion $G_{\mathcal{B}}(H_r, H_l)$ is defined by

$$G_{\mathcal{B}}(H_l, H_r) := G_{0,\mathcal{B}}(H_l, H_r) \boxtimes_{\mathcal{B}(S_m)} H_0(S_m, \mathcal{B}).$$

This construction uses the isomorphism $\mathcal{B}(S_m) \cong \bigoplus_{\lambda\in\Delta} \mathbf{B}(H_\lambda(S_m, \mathcal{B}))$ from Theorem C.9.

LEMMA 4.17. *There is a canonical isomorphism*

$$\overline{H_0(S_m, \mathcal{B})} \boxtimes_{\mathcal{B}(S_m)} H_0(S_m, \mathcal{B}) \cong \mathbb{C}$$

PROOF. The two actions of $\mathcal{B}(S_m)$ factor through its summand $\mathbf{B}(H_0(S_m, \mathcal{B}))$. The result is therefore a special case of the general isomorphism $\bar{H} \boxtimes_{\mathbf{B}(H)} H \cong \mathbb{C}$. □

PROPOSITION 4.18. *Let $\mathcal{B}$ be a conformal net with finite index, and let ${}_{\mathcal{A}}D_{\mathcal{B}}$ and ${}_{\mathcal{B}}E_{\mathcal{C}}$ be irreducible defects. Then there is an isometric embedding*

(4.19) $\Psi : L^2\Big($  $\Big) \to$ 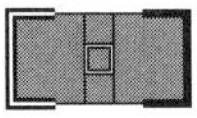 ;

that is, there is a map

$$\Psi = \Psi_{D,E} \colon L^2\big((D \circledast_{\mathcal{B}} E)(S^1_{\top})\big) \to G_{\mathcal{B}}\big(H_0(S_l, D), H_0(S_r, E)\big).$$

As in Theorem 4.11, the map Ψ is equivariant with respect to the left actions of $\{D(I)\}_{I\subset S_{b,\ulcorner}}$, $\{D(I)\}_{I\subset S_{b,\llcorner}}$, $\{E(I)\}_{I\subset S_{b,\urcorner}}$, and $\{E(I)\}_{I\subset S_{b,\lrcorner}}$.

PROOF. By the split property, we can identify  with  $\bar{\otimes}$ , and thus $L^2\big($  $\big)$ with

$L^2\Big($  $\Big) \otimes L^2\Big($  $\Big)$ $= L^2\Big((D \circledast_{\mathcal{B}} E)(S^1_{\top})\Big) \otimes H_0(-S_m, \mathcal{B})$

$$= L^2\Big((D \circledast_{\mathcal{B}} E)(S^1_{\top})\Big) \otimes \overline{H_0(S_m, \mathcal{B})}.$$

Fusing with $H_0(S_m, \mathcal{B})$ and applying Lemma 4.17, we get a canonical isomorphism

$U : L^2\Big($  $\Big) = L^2\Big((D \circledast_{\mathcal{B}} E)(S^1_{\top})\Big)$

$$\xrightarrow{\cong} L^2\Big((D \circledast_{\mathcal{B}} E)(S^1_{\top})\Big) \otimes \overline{H_0(S_m, \mathcal{B})} \underset{\mathcal{B}(S_m)}{\boxtimes} H_0(S_m, \mathcal{B})$$

$\xrightarrow{\cong} L^2\Big($ 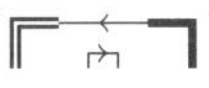 $\Big) \underset{\mathcal{B}(S_m)}{\boxtimes} H_0(S_m, \mathcal{B}).$

Recall from Appendix B.VI that the L^2-space construction is functorial for finite homomorphisms between von Neumann algebras with finite-dimensional center. By Corollary 4.16, the inclusion $\iota\colon$  $\to$ 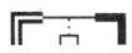 therefore induces a map $L^2(\iota)\colon L^2($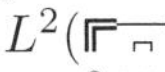$) \to L^2($$)$. Let $L^2(\iota)_{\mathrm{iso}}$ be the isometry in the polar decomposition of $L^2(\iota)$. We set Ψ to be the composite

$\Psi : L^2\Big($  $\Big) \xrightarrow{U} L^2\Big($  $\Big) \boxtimes_{\mathcal{B}(S_m)} H_0(S_m, \mathcal{B})$

(4.20) $\xrightarrow{L^2(\iota)_{\mathrm{iso}}\otimes \mathrm{id}} L^2\Big($ 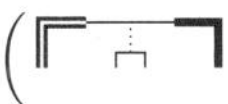 $\Big) \boxtimes_{\mathcal{B}(S_m)} H_0(S_m, \mathcal{B})$

$\xrightarrow{\Psi_0 \otimes \mathrm{id}}$ 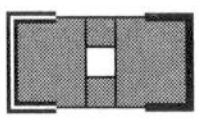 $\underset{\mathcal{B}(S_m)}{\boxtimes} H_0(S_m, \mathcal{B}) = G\big(H_0(D), H_0(E)\big),$

where Ψ_0 is the unitary isomorphism from Theorem 4.11. $\square$

We will prove later, in Theorem 6.2 concerning the composite map (4.31), that the map Ψ is actually an isomorphism. We can already observe the following special case of that result:

LEMMA 4.21. *If D and E are the identity defects of some finite-index conformal net $\mathcal{A}$, then $\Psi_{D,E}$ is a unitary isomorphism.*

PROOF. We need to show that the map

$\big((\Psi_0 \circ L^2(\iota)_{\mathrm{iso}}) \otimes \mathrm{id}\big) \circ U : H_0(S_b) = L^2\Big($  $\Big) \to$ 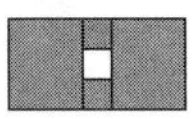 $\underset{\mathcal{A}(S_m)}{\boxtimes} H_0(S_m)$

is an isomorphism. By the computation (4.6), we know that the right-hand side is isomorphic to $H_0(S_b)$, and thus is irreducible as an S_b-sector of $\mathcal{A}$. The above map is a homomorphism of S_b-sectors and is injective by the previous proposition. It is therefore an isomorphism. □

Associativity for the inclusion of the vacuum sector. Using the isometric embedding Ψ from (4.20) in place of the unitary isomorphism Ψ_0 from (4.12), we can form the following diagram analogous to (4.14):

(4.22)

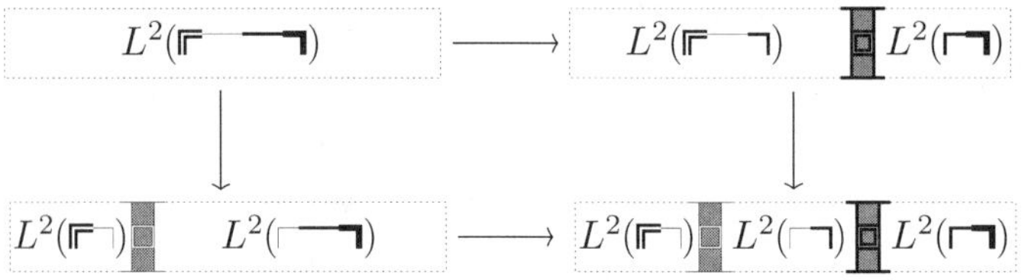

LEMMA 4.23. *Let $\mathcal{B}$ and $\mathcal{C}$ be finite-index conformal nets, and let ${}_{\mathcal{A}}D_{\mathcal{B}}$, ${}_{\mathcal{B}}E_{\mathcal{C}}$, and ${}_{\mathcal{C}}F_{\mathcal{D}}$ be irreducible defects. Then the diagram* (4.22) *is commutative.*

PROOF. By definition, each side of (4.22) is a composite of three maps. Replacing each side by its definition, that diagram can be expanded to a 4×4 grid

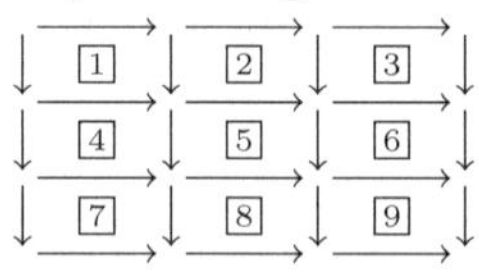

that contains 9 squares. The four upper left squares of that grid are given by

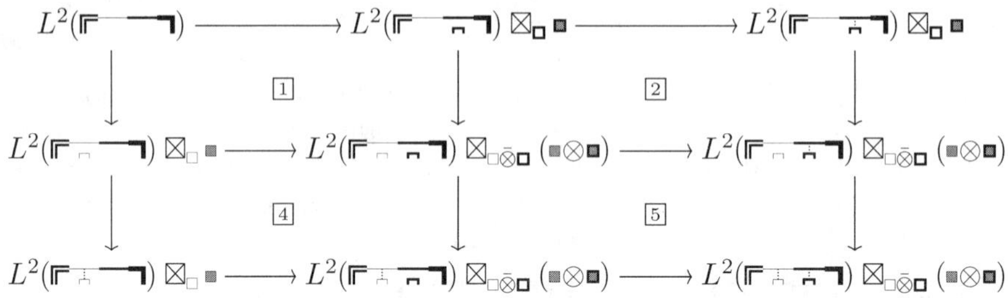

where $\boxtimes_{\square\blacksquare}$ and $\boxtimes_{\square\blacksquare}$ stand for $\boxtimes_{\mathcal{B}(S_m)} H_0(S_m, \mathcal{B})$ and $\boxtimes_{\mathcal{C}(S_m^{\mathrm{tr}})} H_0(S_m^{\mathrm{tr}}, \mathcal{C})$, respectively, and $\boxtimes_{\square\bar{\otimes}\square}(\blacksquare\otimes\blacksquare)$ stands for $\boxtimes_{\mathcal{B}(S_m)\bar{\otimes}\mathcal{C}(S_m^{\mathrm{tr}})}(H_0(S_m, \mathcal{B}) \otimes H_0(S_m^{\mathrm{tr}}, \mathcal{C}))$. Here, S_m^{tr} denotes a translated copy of the circle S_m.

The squares [1], [2], and [4] clearly commute. To see that [5] commutes, note first that ╔ ┴ ┬┐ is a factor, as can be seen by applying Lemma B.36 twice. That square then commutes by the functoriality of $L^2(-)_{\mathrm{iso}}$—see [**BDH11**, Prop. 6.22] and note that the necessary conditions for that functoriality are satisfied by Corollary 4.16 and by Lemma 4.24 below. The upper right squares of our 4×4 gird are given by

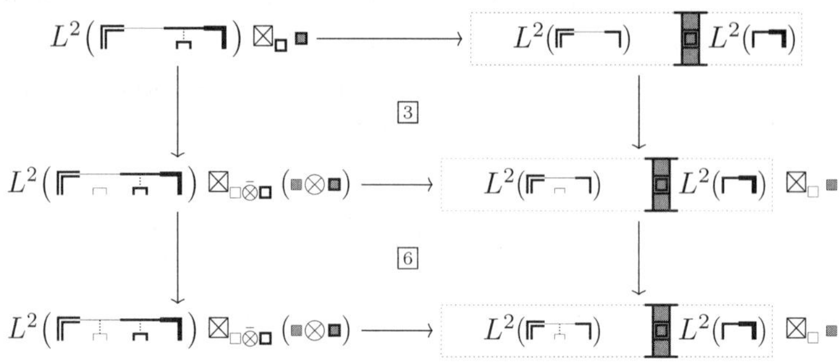

and their commutativity is unproblematic. We refrain from drawing the last row of the gird. The squares $\boxed{7}$ and $\boxed{8}$ are similar to $\boxed{3}$ and $\boxed{6}$. The commutativity of $\boxed{9}$ follows from that of (4.14). □

LEMMA 4.24. *Let ${}_{\mathcal{A}}D_{\mathcal{B}}$, ${}_{\mathcal{B}}E_{\mathcal{C}}$, and ${}_{\mathcal{C}}F_{\mathcal{D}}$ be as in Lemma 4.23. Let* $A :=$, $B :=$, *and let* $\iota : A \to B$ *denote the inclusion. Then* $Z(B) \subset \iota(A)$.

PROOF. By the split property, we have isomorphisms $\cong$ $\bar{\otimes}$ and $\cong$ $\bar{\otimes}$. It is therefore sufficient to show that

Let S_u, K_u, J_u be as in (4.2) and (4.9), and let $H := H_0(S_u, \mathcal{B})$, and $M := \mathcal{B}(K_u)$, with commutant $M' = \hat{\mathcal{B}}(J_u)$. Since H is a faithful M-module, we can pick an M-linear isomorphism $\ell^2 \otimes H \cong \ell^2 \otimes L^2(M)$.[11] Under the corresponding isomorphism of Hilbert spaces $\ell^2 \otimes$ $\cong \ell^2 \otimes$, the algebra

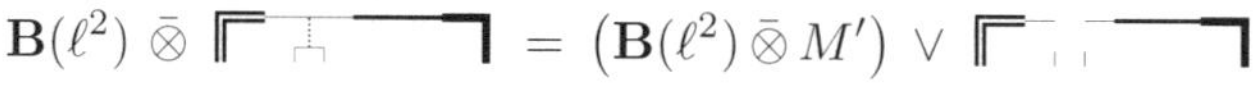

corresponds to

It follows that

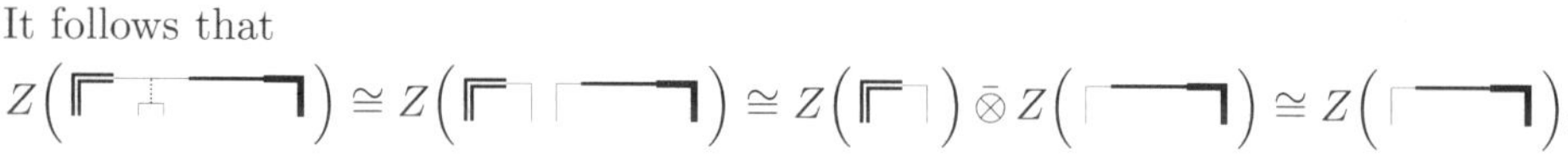

where the last equality is because ${}_{\mathcal{A}}D_{\mathcal{B}}$ is irreducible.

We now argue that the natural inclusion $\hookrightarrow$ induces an isomorphism of centers. By Theorem 3.6, the center of these algebras is finite-dimensional. The center $Z(\;)$ certainly maps to the center $Z(\;)$ and that the map is injective. It is therefore an isomorphism. The claim now follows, as

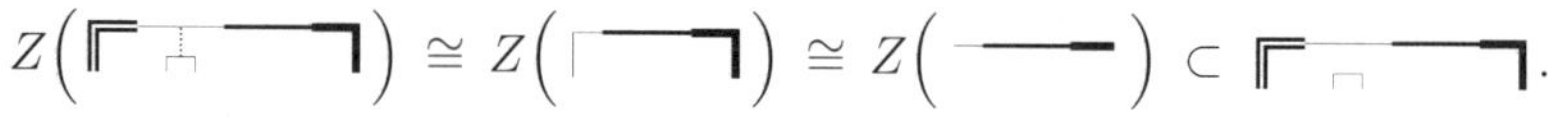

□

REMARK 4.25. All the defects in this section were assumed to be irreducible. However, using the compatibility of directs integrals with various operations, it is straightforward to extend Proposition 4.18 and Lemma 4.23 to arbitrary defects.

4.D. Comparison between fusion and keystone fusion

Let $\mathcal{A}$ be a conformal net with finite index (implicitly irreducible as before). In this section, we will define a unitary natural transformation $\Phi_{\mathcal{A}} : F_{\mathcal{A}} \to G_{\mathcal{A}}$ between the functors introduced in Section 4.A. Graphically, this natural transformation is denoted

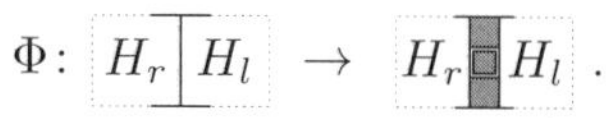

Recall the circles S_l, S_r, and S_b introduced in (4.5):

(4.26) $S_l =$ 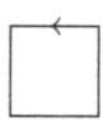 , $S_r =$ 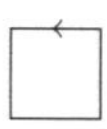, and $S_b =$ 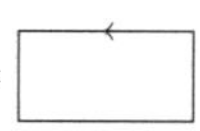 .

As before, we let $I := S_l \cap S_r$, with orientation inherited from S_r. The circles S_l and S_r are given conformal structures by their unit speed parametrizations. The

[11] Here, $\ell^2 := \ell^2(\mathbb{N})$ could be removed from this isomorphism if we knew that M was a type *III* factor, a fact which is likely to be true (unless $\mathcal{B}$ is trivial) but which we haven't proven in our setup.

circle S_b is also given a conformal structure, as follows. Let $j_l \in \mathrm{Conf}_-(S_l)$ and $j_r \in \mathrm{Conf}_-(S_r)$ be the unique conformal involutions fixing ∂I. The conformal structure on S_b is the one making $\epsilon_l := j_l|_I \cup \mathrm{Id}_{j_r(S_r \setminus I)} : S_r \to S_b$ into a conformal map. Equivalently, it is the one for which $\epsilon_r := j_r|_I \cup \mathrm{Id}_{j_l(S_l \setminus I)} : S_l \to S_b$ is a conformal map.

WARNING 4.27. The conformal structure on S_b is *not* the one induced by its constant speed parametrization. Nevertheless, the reflection along the horizontal and vertical symmetry axes of S_b are conformal involutions.

Consider the vacuum sectors $H_0(S_l)$, $H_0(S_r)$, and $H_0(S_b)$ for the net $\mathcal{A}$ on the circles (4.26). By the construction (C.6), there is a canonical identification $\Upsilon : H_0(S_l) \boxtimes_{\mathcal{A}(I)} H_0(S_r) \cong H_0(S_b)$ that is equivariant with respect to the actions of $\mathcal{A}(J)$ for every $J \subset S_b$ (i.e., it is an isomorphism of S_b-sectors of $\mathcal{A}$). We denote it graphically as

(4.28) 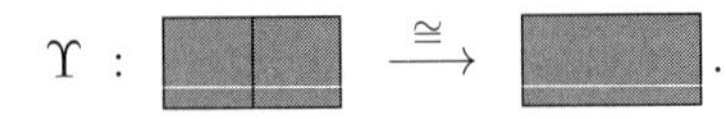

The following proposition improves on Proposition 4.4 by providing an explicit construction of the natural isomorphism $\Phi : F \to G$.

PROPOSITION 4.29. *There is a unitary natural isomorphism $\Phi = \Phi_{\mathcal{A}}$ between the fusion functor $F_{\mathcal{A}}$ and the keystone fusion functor $G_{\mathcal{A}}$:*

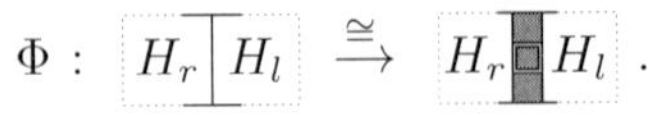

This natural isomorphism is symmetric monoidal in the sense that $\Phi_{\mathcal{A}\otimes\mathcal{B}} = \Phi_{\mathcal{A}} \otimes \Phi_{\mathcal{B}}$.

PROOF. Let I_l and I_r be as in (4.1), and let I_l' and I_r' be the closures of their complements in S_l and S_r, respectively. Since the actions of $\mathcal{A}(I_l)$ and $\mathcal{A}(I_r)$ on $H_0(S_l)$ and $H_0(S_r)$ are faithful, by Lemma 4.3 it is enough to define the isomorphism $\Phi_{H_0(S_l),H_0(S_r)}$: , and to check that it commutes with the natural actions of $\mathcal{A}(I_l)' = \mathcal{A}(I_l')$ and $\mathcal{A}(I_r)' = \mathcal{A}(I_r')$. We define this isomorphism as the composite

(4.30) 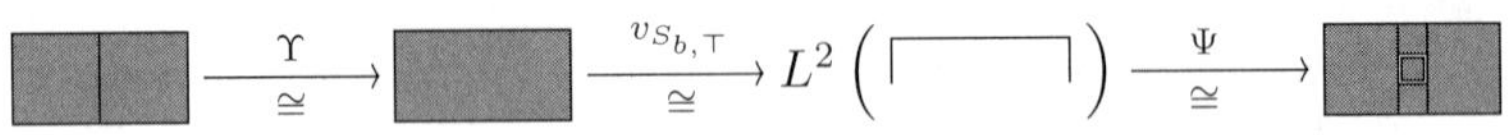

where $v_{S_{b,\top}} : H_0(S_b, \mathcal{A}) \to L^2(\mathcal{A}(S_{b,\top}))$ is the canonical unitary isomorphism (C.2) associated to the upper half $S_{b,\top}$ of the conformal circle S_b, and Ψ is the unitary isomorphism from Lemma 4.21. The symmetric monoidal condition is clear by construction. $\square$

Let $\mathcal{A}$ and $\mathcal{C}$ be conformal nets, let $\mathcal{B}$ be a conformal net with finite index, and let ${}_{\mathcal{A}}D_{\mathcal{B}}$ and ${}_{\mathcal{B}}E_{\mathcal{C}}$ be defects. Let us introduce the notation  for the Hilbert space $L^2($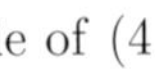$) = L^2((D \circledast_{\mathcal{B}} E)(S^1_\top))$ that appears in the left-hand side of (4.19). Combining Proposition 4.18 (see also Remark 4.25) and Proposition 4.29, we can construct an isometric map

(4.31) 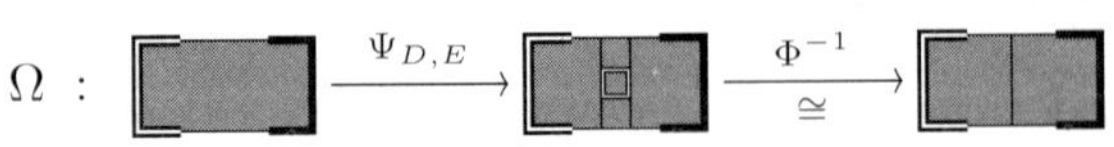

where Φ stands for $\Phi_{H_0(S_l,D),H_0(S_r,E)}$. We will show later, in Theorem 6.2, that the map $\Omega = \Omega_{D,E}$ is in fact an isomorphism. This map is the fundamental "$1 \boxtimes 1 = 1$ isomorphism" comparing $1_{D \circledast E}$ with $1_D \boxtimes 1_E$.

PROPOSITION 4.32. *Let $\mathcal{A}$, $\mathcal{B}$, $\mathcal{C}$, $\mathcal{D}$ be conformal nets, of which the second and third are assumed to be finite, and let ${}_{\mathcal{A}}D_{\mathcal{B}}$, ${}_{\mathcal{B}}E_{\mathcal{C}}$, ${}_{\mathcal{C}}F_{\mathcal{D}}$ be defects. Then the maps*

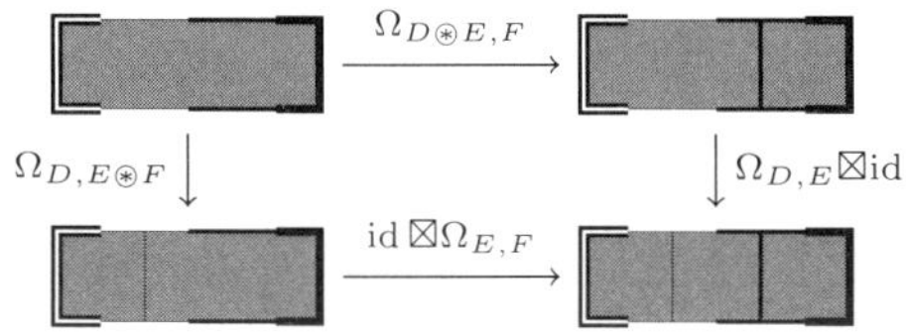

form a commutative diagram.

PROOF. By the definition of Ω, the above diagram can be expanded to

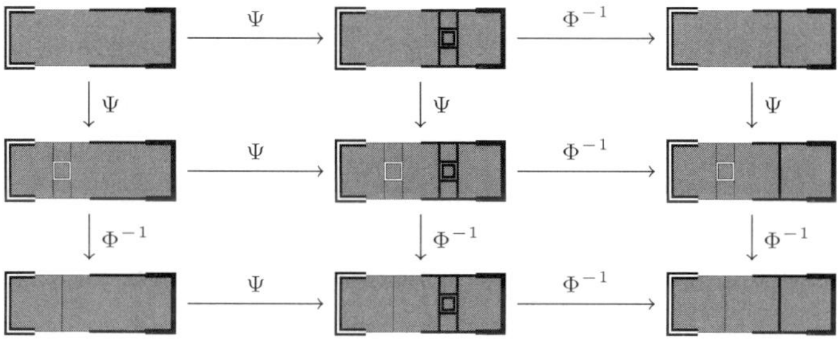

The upper left square commutes by Lemma 4.23 (see also Remark 4.25). The remaining three squares commute by the naturality of Φ^{-1}. □

CHAPTER 5

Haag duality for composition of defects

Throughout this chapter we fix conformal nets $\mathcal{A}$, $\mathcal{B}$, and $\mathcal{C}$, always assumed to be irreducible, and irreducible defects ${}_{\mathcal{A}}D_{\mathcal{B}}$ and ${}_{\mathcal{B}}E_{\mathcal{C}}$. In our pictures, we will use the notation ═ for intervals on which we evaluate $\mathcal{A}$, we will use ─ for intervals on which we evaluate $\mathcal{B}$, and ▬ for intervals on which we evaluate $\mathcal{C}$. We will also use ═─ for bicolored intervals on which we evaluate D, and ─▬ for bicolored intervals on which we evaluate E.

Let S_l and S_r be as in (4.5) and (4.7), with intersection I oriented like S_r. As before, we use the notation ▯ $:= H_0(S_l, D) = L^2(⌜⌝)$, similarly ▯ $:= H_0(S_r, E) = L^2(⌜⌝)$, and ▯▯ $:= H_0(S_l, D) \boxtimes_{\mathcal{B}(I)} H_0(S_r, E)$. We will again be using the Notation 1.49. Letting

$$
(5.1)\qquad
\begin{array}{lll}
I_1 := \partial^{\ulcorner}([0,1]\times[\tfrac{1}{2},1]) & (I_1)_\circ := (I_1)_{x\le\frac{1}{2}} & (I_1)_\bullet := (I_1)_{x\ge\frac{1}{2}} \\
I_2 := \partial^{\urcorner}([1,2]\times[\tfrac{1}{2},1]) & (I_2)_\circ := (I_2)_{x\le\frac{3}{2}} & (I_2)_\bullet := (I_2)_{x\ge\frac{3}{2}} \\
I_3 := \partial^{\llcorner}([0,1]\times[0,\tfrac{1}{2}]) & (I_3)_\circ := (I_3)_{x\le\frac{1}{2}} & (I_3)_\bullet := (I_3)_{x\ge\frac{1}{2}} \\
I_4 := \partial^{\lrcorner}([1,2]\times[\tfrac{1}{2},1]) & (I_4)_\circ := (I_4)_{x\le\frac{3}{2}} & (I_4)_\bullet := (I_4)_{x\ge\frac{3}{2}},
\end{array}
$$

we will write $D(⌜)$, $E(⌝)$, $D(⌞)$, $E(⌟)$ for $D(I_1)$, $E(I_2)$, $D(I_3)$, $E(I_4)$, respectively.

Main Theorem 5.2. *Assuming $\mathcal{B}$ has finite index, then on the Hilbert space* ▯▯, *we have*

$$
(5.3)\qquad D(⌜)\vee E(⌝) = \big(D(⌞)\vee E(⌟)\big)'.
$$

Proof. Let us introduce some notation for various algebras that act on the Hilbert space ▯▯. The main algebras of interest are ⌜ ⌝ $= (D \circledast_{\mathcal{B}} E)(S^1_\top)$ and ⌞ ⌟ $= (D \circledast_{\mathcal{B}} E)(S^1_\perp)$, and our goal is to show that the inclusion

$$
(5.4)\qquad ⌜\;⌝ \subseteq \big(⌞\;⌟\big)'
$$

is an isomorphism. Let us fix once and for all a small number ϵ. Consider the 1-manifolds

$$
\begin{aligned}
J_0 &:= \partial^{\ulcorner}\big([0,\tfrac{1}{2}+\epsilon]\times[\tfrac{1}{2},1]\big),\\
J_1 &:= \big([\tfrac{1}{2}+\epsilon,\tfrac{2}{3}]\cup[\tfrac{5}{6},1]\big)\times\{1\},\\
J_2 &:= \big([1,\tfrac{3}{2}-\epsilon]\times\{1\}\big)\cup\big([\tfrac{7}{6},\tfrac{4}{3}]\times\{0\}\big),\\
J_3 &:= \partial^{\urcorner}\big([\tfrac{3}{2}-\epsilon,2]\times[\tfrac{1}{2},1]\big).
\end{aligned}
$$

We will use the following algebras:

(5.5)  $:= D(J_0) \vee \mathcal{B}(J_1) \vee \mathcal{B}(J_2) \vee E(J_3) = D(J_0 \cup J_1) \vee E(J_2 \cup J_3)$

(5.6) 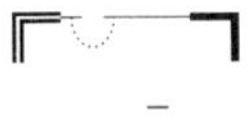 $:= D(J_0) \vee \hat{\mathcal{B}}(J_1) \vee \mathcal{B}(J_2) \vee E(J_3) = \hat{D}(J_0 \cup J_1) \vee E(J_2 \cup J_3)$

(5.7) 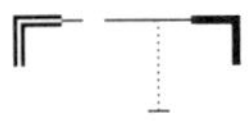 $:= D(J_0) \vee \mathcal{B}(J_1) \vee \hat{\mathcal{B}}(J_2) \vee E(J_3) = D(J_0 \cup J_1) \vee \hat{E}(J_2 \cup J_3)$

(5.8) 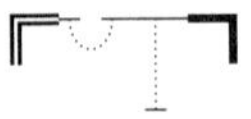 $:= D(J_0) \vee \hat{\mathcal{B}}(J_1) \vee \hat{\mathcal{B}}(J_2) \vee E(J_3) = \hat{D}(J_0 \cup J_1) \vee \hat{E}(J_2 \cup J_3).$

Here $\hat{\mathcal{B}}$, $\hat{D}$, and $\hat{E}$ are as in 3.11, and J_0 and J_3 are bicolored as in (4.7). By Lemma 3.12, the algebras $\hat{\mathcal{B}}(J_1)$ and $\hat{\mathcal{B}}(J_2)$ act on  and  respectively, and satisfy $D(J_0) \vee \hat{\mathcal{B}}(J_1) = \hat{D}(J_0 \cup J_1)$ and $\hat{\mathcal{B}}(J_2) \vee E(J_3) = \hat{E}(J_2 \cup J_3)$. The equalities in (5.6)–(5.8) follow.

In Section 5.A below we will obtain some purchase on the Haag inclusion 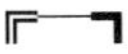 $\subseteq$ (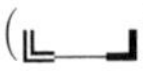)$'$ by showing that its statistical dimension is the same as that of the inclusion  $\subseteq$ ()$'$. (Here the algebra  is defined similarly to 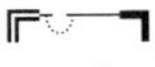.) We can compute the statistical dimension of that latter inclusion by squeezing it into a sequence of simpler inclusions of von Neumann algebras, as follows:

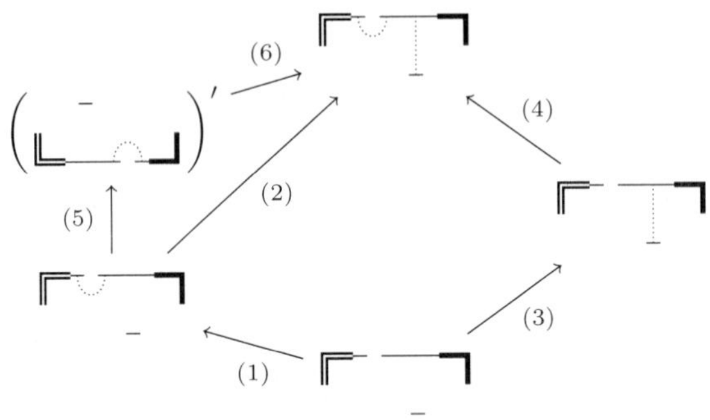

Because D and E are irreducible, the algebra (5.5) is a factor. Using Lemma 3.12, note that the algebra  (the right connected component of the picture (5.7)) is the commutant of a factor acting on a vacuum sector for E; it follows that (5.7) is a factor. More difficult is the fact that (5.8) is a factor—that is the content of Corollary 5.16, following from Lemma 5.13 below. The algebra (5.6) is not a factor, but combining Lemma 5.10 below and Theorem 3.6, we will learn that it does have finite-dimensional center; let n be the dimension of this center.

Let $\nu_1, \ldots, \nu_6$ be the matrices of statistical dimensions (Appendix B.VII) of the various inclusions in the above diagram:

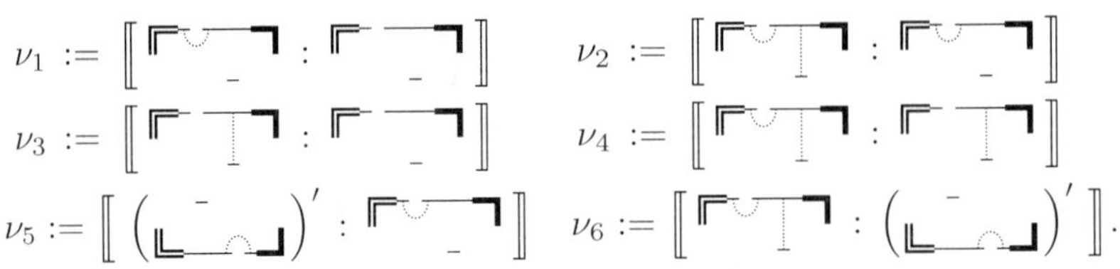

Note that ν_3 and ν_4 are scalars, ν_1 is a row vector, ν_2 and ν_6 are column vectors, and ν_5 is an $n \times n$ matrix. As the matrix of statistical dimensions is multiplicative (B.17), it follows that $\nu_1 \nu_2 = \nu_3 \nu_4$, and that $\nu_2 = \nu_5 \nu_6$.

We will need the following facts about $\nu_1, \ldots, \nu_6$:

(i) There is an equality of $n \times n$ matrices

$\nu_5 \equiv$ 

This is proven in Lemma 5.10 below.

(ii) The map (4.31) exhibits $L^2($$)$ as a sub-bimodule of 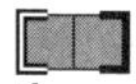. Using the additivity of statistical dimension (B.13), we obtain the *entrywise* matrix inequality

$\nu_5 =$ 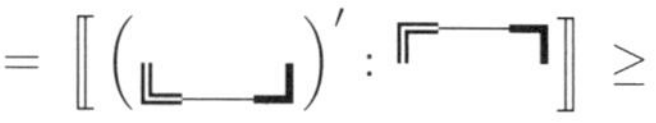 $\geq \mathbf{1}_n ,$

where $\mathbf{1}_n$ denotes the identity matrix.

(iii) In Corollary 5.17, we will show that ν_1 is the transpose of ν_6.

(iv) In Corollary 5.21, we will show that $\nu_3 = \sqrt{\mu}$ and $\nu_4 = \sqrt{\mu}$, where $\mu = \mu(\mathcal{B})$ is the index of the conformal net $\mathcal{B}$.

Using these results and the fact that the matrices of statistical dimensions have only non-negative entries, we can now compute that

$$\| \nu_2 \|^2 = \nu_2^T \nu_2 = \nu_6^T \nu_5^T \nu_5 \nu_6 = \nu_1 \nu_5^T \nu_5 \nu_6 \geq \nu_1 \nu_5 \nu_6 = \nu_3 \nu_4 = \mu ,$$

with equality if and only if $\nu_5 = \mathbf{1}_n$ (here, ^{T} denotes transpose). However, by applying (B.20) to the algebras $M := \hat{\mathcal{B}}(J_2)$, $N := \mathcal{B}(J_2)$, and $A := D(J_0) \vee \hat{\mathcal{B}}(J_1) \vee E(J_3)$, we obtain the reverse inequality

$$\|\nu_2\| \leq \sqrt{\mu}.$$

It follows that $\nu_5 = \mathbf{1}_n$, and the inclusion (5.4) is therefore an isomorphism. □

In Theorem 5.2, the defects D and E were assumed to be irreducible, but the statement holds in general:

Corollary 5.9. *Let ${}_{\mathcal{A}}D_{\mathcal{B}}$ and ${}_{\mathcal{B}}E_{\mathcal{C}}$ be defects. If the conformal net $\mathcal{B}$ has finite index, then the algebra $D($$) \vee E($$)$ is the commutant of $D($$) \vee E($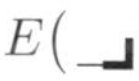$)$ on* .

Proof. We need to show that $D(I_1) \vee E(I_2) = (D(I_3) \vee E(I_4))'$, where the intervals I_1, I_2, I_3, I_4 are as in (5.1). Disintegrating

$$D = \int^{\oplus} D_x \qquad \text{and} \qquad E = \int^{\oplus} E_y$$

into irreducible defects, the Hilbert space 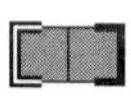 $= H_0(S_l, D) \boxtimes_{\mathcal{B}(I)} H_0(S_r, E)$ decomposes correspondingly as $\iint^{\oplus} H_0(S_l, D_x) \boxtimes_{\mathcal{B}(I)} H_0(S_r, E_y)$. This induces direct integral decompositions

$$D(I_1) \vee E(I_2) = \iint^{\oplus} D_x(I_1) \vee E_y(I_2),$$
$$D(I_3) \vee E(I_4) = \iint^{\oplus} D_x(I_3) \vee E_y(I_4),$$

and therefore also $(D(I_3) \vee E(I_4))' = \iint^{\oplus} (D_x(I_3) \vee E_y(I_4))'$, where the commutant of $D_x(I_3) \vee E_y(I_4)$ is taken on $H_0(S_l, D_x) \boxtimes_{\mathcal{B}(I)} H_0(S_r, E_y)$. By Theorem 5.2, we have $D_x(I_1) \vee E_y(I_2) = (D_x(I_3) \vee E_y(I_4))'$, and the result follows. □

5.A. The dimension of the Haag inclusion

Recall the notion of the center of a defect from Section 1.C: for a genuinely bicolored interval I, the algebra $Z(D(I))$ is independent of I (up to canonical isomorphism), and is denoted $Z(D)$.

From now on, the defects D and E are again assumed irreducible:

LEMMA 5.10. *Let X be the set of irreducible summands of $D \circledast_{\mathcal{B}} E$. We have a canonical identification of centers $Z(D \circledast_{\mathcal{B}} E) = Z($$) = Z($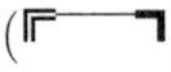$) = Z($$) = Z($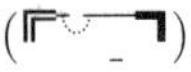$)$, and we have the following equality of $X \times X$ matrices:*

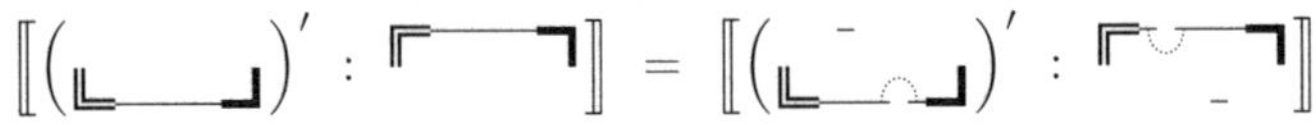

PROOF. Note that $[\![A : B]\!] = [\![A \,\bar{\otimes}\, C : B \,\bar{\otimes}\, C]\!]$ whenever C is a factor (B.18). The algebras $\mathcal{B}([\frac{7}{6}, \frac{4}{3}])$ and $\mathcal{B}([\frac{7}{6} - \epsilon, \frac{4}{3} + \epsilon])'$ are split on $H_0(S_r, \mathcal{B})$, and hence on any $\mathcal{B}([\frac{7}{6} - \epsilon, \frac{4}{3} + \epsilon])$-module. Since $($$)' \subset \mathcal{B}([\frac{7}{6} - \epsilon, \frac{4}{3} + \epsilon])'$ on 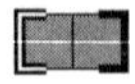, it follows that

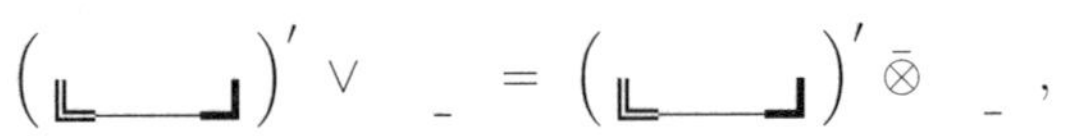

where the line stands for $\mathcal{B}([\frac{7}{6}, \frac{4}{3}] \times \{0\})$. We conclude that

(5.11)

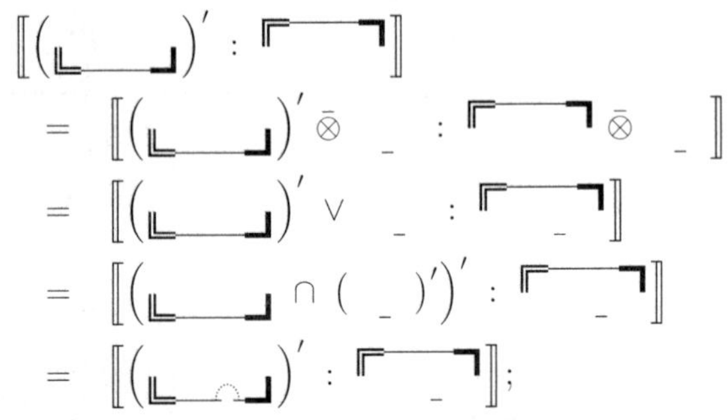

the last equality follows by applying Lemma B.26 with $B :=$ , $A :=$, $A_0 :=$, and $A \cap A_0' =$. Similarly, we have

(5.12)

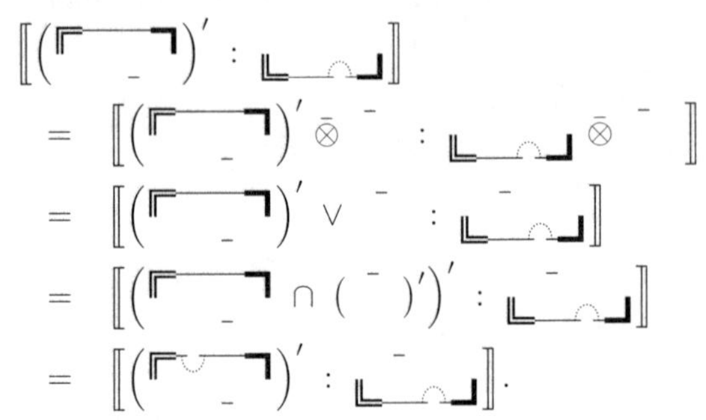

Taking commutants transposes the matrix of statistical dimensions (B.16). Thus (5.12) implies

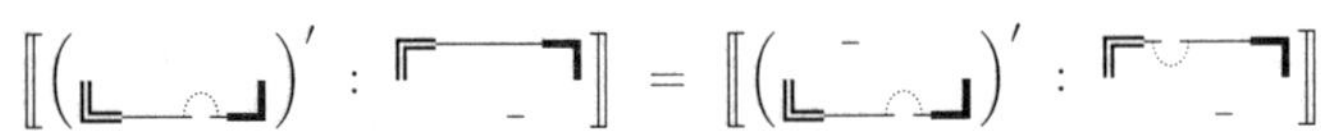

which, combined with (5.11), proves the Lemma. □

5.B. The double bridge algebra is a factor

Let S_l and S_r be as in (4.5) and now let $\tilde{S}_l := \partial([0,\frac{5}{4}]\times[0,1])$ and $\tilde{S}_r := \partial([\frac{5}{4},2]\times[0,1])$; give these circles the bicoloring

$$(S_l)_\circ := (S_l)_{x\le\frac{1}{2}} \quad (S_l)_\bullet := (S_l)_{x\ge\frac{1}{2}} \quad (S_r)_\circ := (S_r)_{x\le\frac{3}{2}} \quad (S_r)_\bullet := (S_r)_{x\ge\frac{3}{2}}$$
$$(\tilde{S}_l)_\circ := (\tilde{S}_l)_{x\le\frac{1}{2}} \quad (\tilde{S}_l)_\bullet := (\tilde{S}_l)_{x\ge\frac{1}{2}} \quad (\tilde{S}_r)_\circ := (\tilde{S}_r)_{x\le\frac{3}{2}} \quad (\tilde{S}_r)_\bullet := (\tilde{S}_r)_{x\ge\frac{3}{2}}.$$

Let $I := S_l\cap S_r$ and $\tilde{I} := \tilde{S}_l\cap\tilde{S}_r$. Recall from (C.2) that for any conformal circle S and any interval $I\subset S$, the vacuum sector $H_0(S,\mathcal{B})$ is a unit for Connes fusion over $\mathcal{B}(I)$. By applying this fact twice, we can construct a non-canonical isomorphism

$$H_0(S_l,D)\circledast_{\mathcal{B}(I)} H_0(S_r,E) \cong H_0(\tilde{S}_l,D)\circledast_{\mathcal{B}(\tilde{I})} H_0(\tilde{S}_r,E),$$

equivariant with respect to the actions of $D(J)$ for every $J\subset$  and $J\subset$ 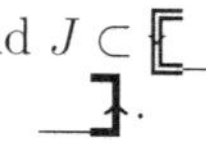, and with respect to the actions of $E(J)$ for every $J\subset$  and $J\subset$ .

Recall the fiber product operation $*$ from Appendix B.IV. Let

$$J := [\tfrac{7}{6},\tfrac{4}{3}]\times\{0,1\},\quad K_l := \partial^{\sqsupset}([\tfrac{7}{6},\tfrac{5}{4}]\times[0,1]),\quad K_r := \partial^{\sqsubset}([\tfrac{5}{4},\tfrac{4}{3}]\times[0,1]),$$
$$K_l' := \partial^{\sqsubset}([0,\tfrac{7}{6}]\times[0,1]),\qquad K_r' := \partial^{\sqsupset}([\tfrac{4}{3},2]\times[0,1]),$$

which we draw for convenience:

$J =$ , $K_l =$, $K_r =$, $K_l' =$ , $K_r' =$ .

We then have $\tilde{S}_l = K_l\cup K_l'$ and $\tilde{S}_r = K_r\cup K_r'$. We use $H = H_0(\mathcal{B},\tilde{S}_l)$ and $K = H_0(\mathcal{B},\tilde{S}_r)$ in the definition B.10 of the fiber product $\mathcal{B}(K_l)*_{\mathcal{B}(\tilde{I})}\mathcal{B}(K_r) = \big(\mathcal{B}(K_l)'\vee\mathcal{B}(K_r)'\big)'$. By Haag duality, we have $\mathcal{B}(K_l)' = \mathcal{B}(K_l')$ and $\mathcal{B}(K_r)' = \mathcal{B}(K_r')$. These algebras all act on $H_0(\mathcal{B},\tilde{S}_l)\boxtimes_{\mathcal{B}(\tilde{I})} H_0(\mathcal{B},\tilde{S}_r)$, which can be identified with $H_0(\mathcal{B};S_b)$ by (C.5). Altogether we obtain

$$\hat{\mathcal{B}}(J) = \big(\mathcal{B}(K_l')\vee\mathcal{B}(K_r')\big)' = \big(\mathcal{B}(K_l)'\vee\mathcal{B}(K_r)'\big)' = \mathcal{B}(K_l)*_{\mathcal{B}(\tilde{I})}\mathcal{B}(K_r).$$

We denote the above equation graphically by 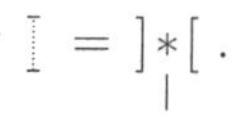.

LEMMA 5.13. *We have the following equality of subalgebras of* $\mathbf{B}($$)$:

(5.14) 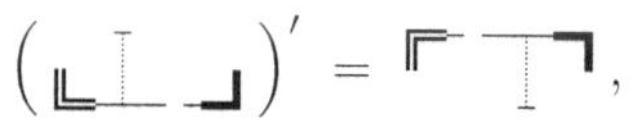

(5.15) 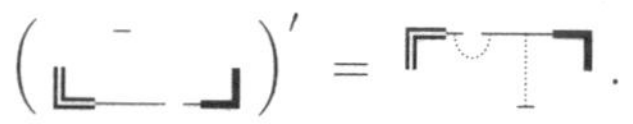

PROOF. By Lemma 3.12 and Proposition 1.16, respectively, we have

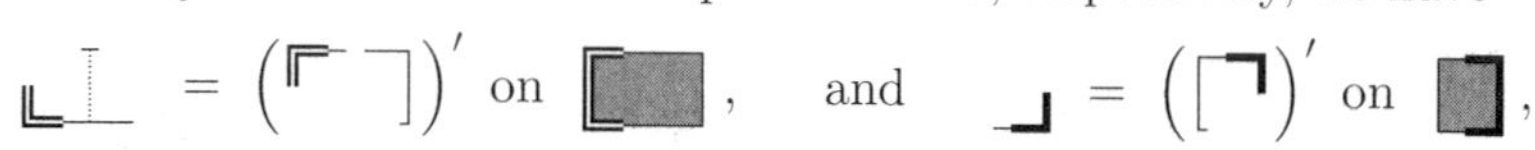

where 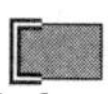 stands for $H_0(\tilde{S}_l,D)$, and  stands for $H_0(\tilde{S}_r,E)$.

We have the following sequence of equalities

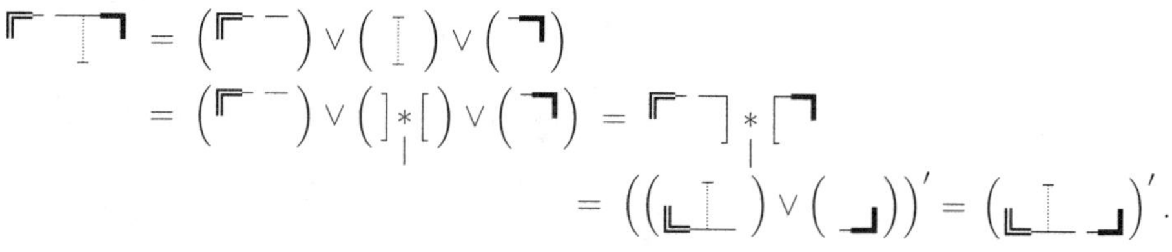

Here the third equality uses Lemma B.32. By Lemma 3.12, we also have

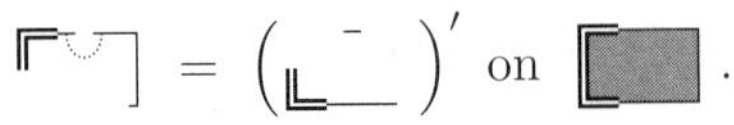

We therefore similarly have

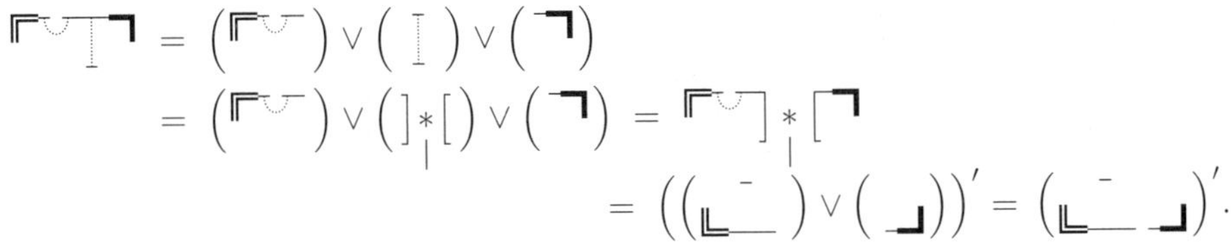

□

COROLLARY 5.16. *The algebra* 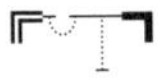 *is a factor.* □

COROLLARY 5.17. *We have*

(5.18) 

PROOF. By (B.16) and by Lemma 5.13, the right-hand side of (5.18) is equal to

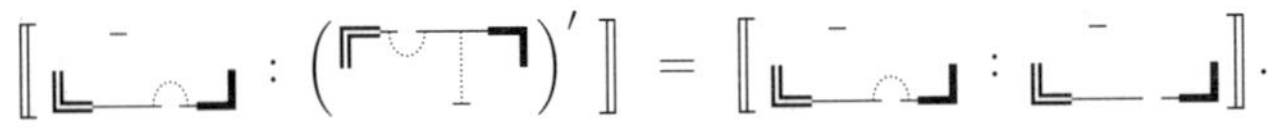

The algebras  and 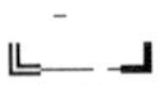 are related to those on the left-hand side of (5.18) by the action of orientation reversing diffeomorphisms of the underlying 1-manifolds: these diffeomorphisms induce algebra isomorphisms  $\cong$  and 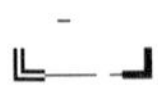 $\cong$ . The result now follows since $[\![A : B]\!] = [\![A^{op} : B^{op}]\!]$. □

5.C. The dimension of the bridge inclusions

LEMMA 5.19. *We have the following equalities of statistical dimensions:*

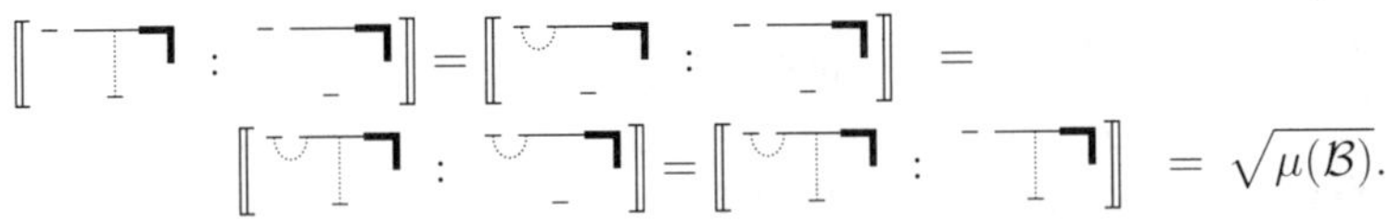

PROOF. We have

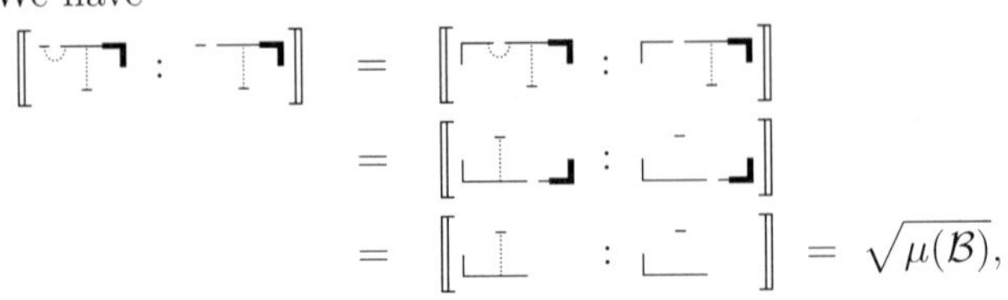

where the first equality is obtained by using an appropriate diffeomorphism, the second one follows from (B.16) and the special case of Lemma 5.13 when D is an identity defect, and the third one uses (B.18).

Let us introduce the auxiliary quantity

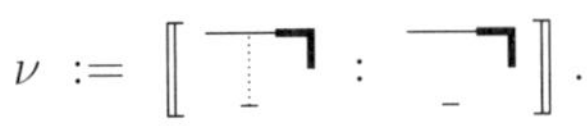

By Lemma 3.15, we know that $\nu \leq \sqrt{\mu(\mathcal{B})}$; in particular $\nu < \infty$. By (B.18), we have

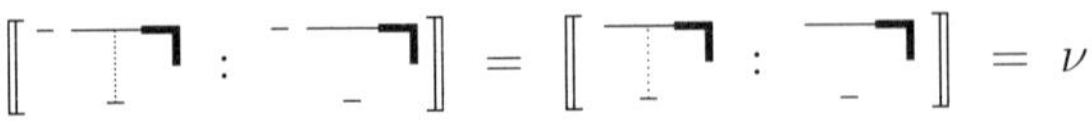

and

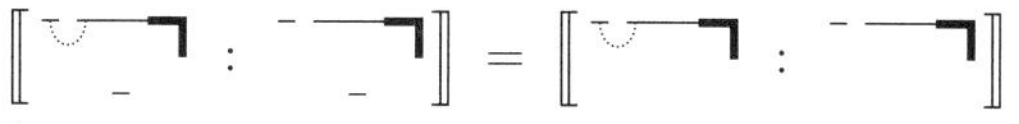 $= \nu$.

Next observe that

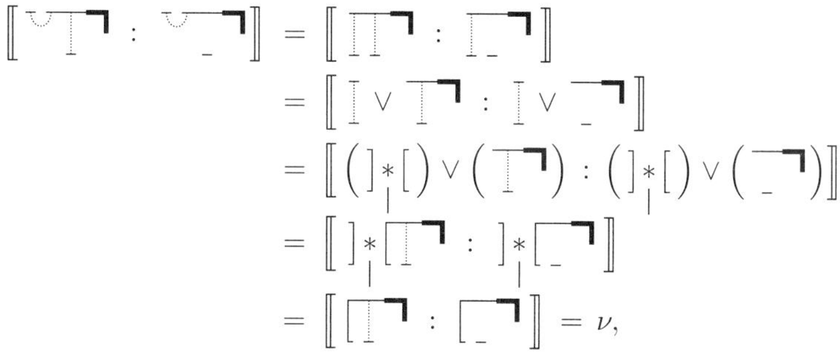

where the fourth and fifth equalities follow from Lemmas B.32 and B.30, respectively.

We conclude the argument by noting that, by (B.17),

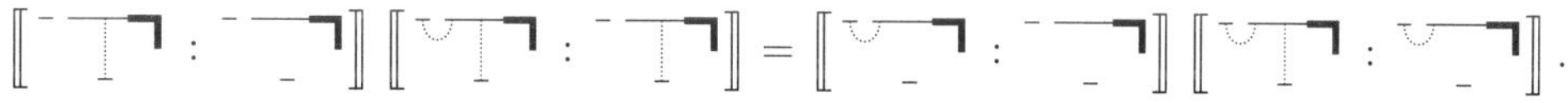

In light of the above computations, that equation gives $\nu\sqrt{\mu(\mathcal{B})} = \nu^2$; since ν is finite, we must have $\nu = \sqrt{\mu(\mathcal{B})}$, as required. □

As a corollary, we obtain the following improvement on Lemma 3.15:

COROLLARY 5.20. *We have*

 $= \sqrt{\mu(\mathcal{B})}$. □

COROLLARY 5.21. *We have the following two equalities:*

 $= \sqrt{\mu(\mathcal{B})}$,

 $= \sqrt{\mu(\mathcal{B})}$.

PROOF. The first equality follows immediately from Corollary 5.20. For the second equality, note that 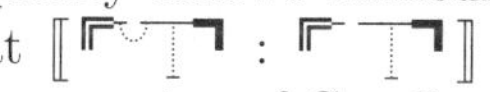 $=$ 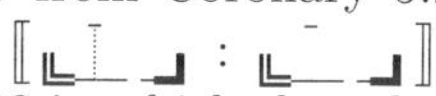 by Lemma 5.13; the result follows by a version of Corollary 5.20 in which the roles of the nets $\mathcal{A}$ and $\mathcal{C}$ have been interchanged. □

CHAPTER 6

The $1 \boxtimes 1$-isomorphism

We are now in a position to prove that the map Ω (4.31), from the vacuum sector of the composition of two defects to the fusion of the vacuum sectors of the individual defects, is an isomorphism. This isomorphism provides the modification

(6.1)

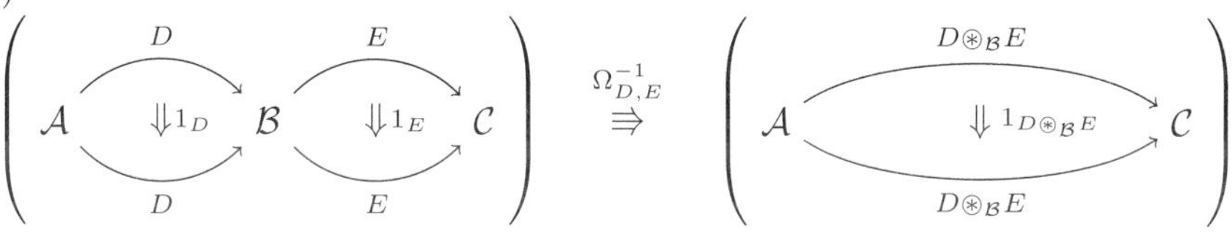

that one expects in any 3-category. More importantly, it also provides the foundation for our construction of the fundamental interchange modification

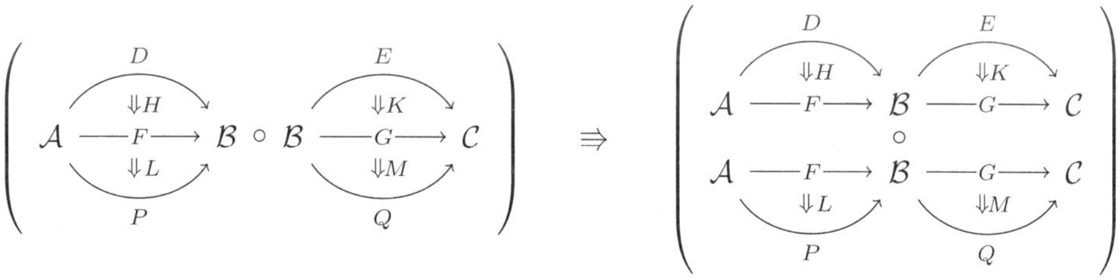

present in any 3-category; see Section 6.D.

6.A. The $1 \boxtimes 1$-map is an isomorphism

Let $\mathcal{A}$, $\mathcal{B}$, and $\mathcal{C}$ be conformal nets, always assumed to be irreducible, and let ${}_{\mathcal{A}}D_{\mathcal{B}}$ and ${}_{\mathcal{B}}E_{\mathcal{C}}$ be defects. Assume furthermore that $\mathcal{B}$ has finite index. As before, we let 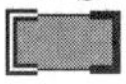 denote the Hilbert space $L^2((D \circledast E)(S^1_\top))$, and we let 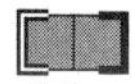 denote the fusion $L^2(D(S^1_\top)) \boxtimes_{\mathcal{B}(I)} L^2(E(S^1_\top))$, where I is the middle vertical interval as in (4.1).

MAIN THEOREM 6.2. *Let $\mathcal{A}$, $\mathcal{B}$, and $\mathcal{C}$ be irreducible conformal nets, and let ${}_{\mathcal{A}}D_{\mathcal{B}}$ and ${}_{\mathcal{B}}E_{\mathcal{C}}$ be defects. The map*

(6.3) $\Omega_{D,E}$: 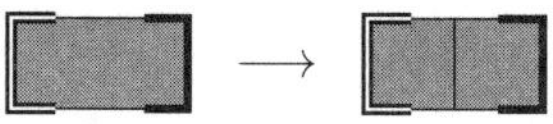

constructed in (4.31) *is a unitary isomorphism.*

PROOF. Because the constructions of the source and target of $\Omega_{D,E}$ commute with direct integrals, we may assume without loss of generality that D and E are irreducible. By construction the map $\Omega_{D,E}$ is an isometry. The algebras  $= (D \circledast E)(S^1_\top)$ and  $= (D \circledast E)(S^1_\perp)$ act faithfully on both sides of (6.3). They are certainly each other's commutants on 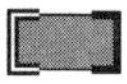, and the

$(D \circledast E)(S^1_\top)$ - $(D \circledast E)(S^1_\perp)$-bimodule has the identity matrix as its matrix of statistical dimensions. We already know (by Propositions 4.18 and 4.29) that $\Omega_{D,E}$ is an embedding. By Theorem 5.2, the algebras and are also each other's commutants on . It follows (since statistical dimension is additive and every nonzero bimodule has nonzero statistical dimension) that the $(D \circledast E)(S^1_\top)$ - $(D \circledast E)(S^1_\perp)$-bimodule also has the identity matrix as its matrix of statistical dimensions, and that $\Omega_{D,E}$ is an isomorphism. □

Given the crucial importance of the "1 times 1 isomorphism" Ω, we collect in one place the main ingredients used in its definition. These are the unitary isomorphisms Ψ from (4.20) and Φ from (4.30); the $1 \boxtimes 1$-isomorphism is the composite Ω from (4.31):

Definition of the unitary Ω : $\xrightarrow{\cong}$, *quick summary:*

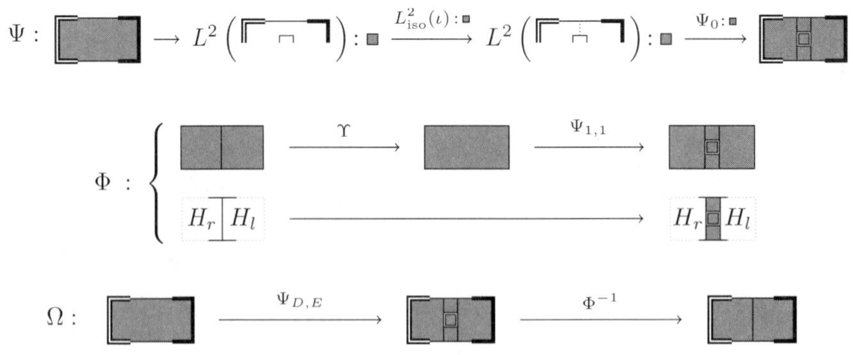

Here, the symbol ": ■" is an abbreviation for $\boxtimes_{\square\blacksquare}$, that is $\boxtimes_{\mathcal{B}(S_m)} H_0(S_m, \mathcal{B})$.

The $1 \boxtimes 1$ isomorphism is monoidal. We now address the compatibility of the isomorphism Ω with the symmetric monoidal structure on conformal nets and defects. Let ${}_{\mathcal{A}_1}(D_1)_{\mathcal{B}_1}$, ${}_{\mathcal{B}_1}(E_1)_{\mathcal{C}_1}$, ${}_{\mathcal{A}_2}(D_2)_{\mathcal{B}_2}$ and ${}_{\mathcal{B}_2}(E_2)_{\mathcal{C}_2}$ be defects. We abbreviate $D_1 := D_1(S^1_\top)$, $D_2 := D_2(S^1_\top)$, $D_1 \otimes D_2 := (D_1 \otimes D_2)(S^1_\top)$, $E_1 := E_1(S^1_\top)$, $E_2 := E_2(S^1_\top)$, $E_1 \otimes E_2 := (E_1 \otimes E_2)(S^1_\top)$, and $B_1 := \mathcal{B}_1(I)$, $B_2 := \mathcal{B}_2(I)$, $B_1 \otimes B_2 := (\mathcal{B}_1 \otimes \mathcal{B}_2)(I)$, where as before I is the vertical unit interval. Combining the various isomorphisms from Appendix B.V we have canonical isomorphisms

$$\begin{aligned}(6.4)\quad & L^2(D_1 \circledast_{B_1} E_1) \otimes L^2(D_2 \circledast_{B_2} E_2) \\ & \qquad \cong \quad L^2((D_1 \otimes D_2) \circledast_{B_1 \otimes B_2} (E_1 \otimes E_2)), \\ (6.5)\quad & (L^2(D_1) \boxtimes_{B_1} L^2(E_1)) \otimes (L^2(D_2) \boxtimes_{B_2} L^2(E_2)) \\ & \qquad \cong \quad L^2(D_1 \otimes D_2) \boxtimes_{B_1 \otimes B_2} L^2(E_1 \otimes E_2).\end{aligned}$$

PROPOSITION 6.6. *The unitary isomorphism* Ω *is compatible with the symmetric monoidal structure on conformal nets and defects. More precisely, if* $\mathcal{B}_1$ *and* $\mathcal{B}_2$ *have finite index then*

$$(6.5) \circ (\Omega_{D_1,E_1} \otimes \Omega_{D_2,E_2}) = \Omega_{D_1 \otimes D_2, E_1 \otimes E_2} \circ (6.4). \tag{6.7}$$

PROOF. For every step in the construction of Ω there is an isomorphism analogous to (6.5) or (6.4). It is a lengthy, but not difficult, exercise to check that for each step in the construction of Ω the corresponding version of equation (6.7) holds. □

WARNING 6.8. *It appears that Ω is not a natural transformation!* More precisely, there seem to exist irreducible defects ${}_{\mathcal{A}}D_{\mathcal{B}}$, ${}_{\mathcal{B}}E_{\mathcal{C}}$, ${}_{\mathcal{A}}F_{\mathcal{B}}$, ${}_{\mathcal{B}}G_{\mathcal{C}}$ and finite natural transformations $\tau : D \to F$ and $\sigma : E \to G$ (see Definition 1.37) for which the following diagram fails to be commutative:

$$\begin{array}{ccc} H_0(D \circledast_{\mathcal{B}} E) & \xrightarrow{\Omega_{D,E}} & H_0(D) \boxtimes_{\mathcal{B}} H_0(E) \\ {\scriptstyle H_0(\tau\circledast\sigma)}\downarrow & & \downarrow{\scriptstyle H_0(\tau)\boxtimes H_0(\sigma)} \\ H_0(F \circledast_{\mathcal{B}} G) & \xrightarrow{\Omega_{F,G}} & H_0(F) \boxtimes_{\mathcal{B}} H_0(G) \end{array} \tag{6.9}$$

Here, $H_0(\tau)$ is the value of the functor L^2 on the map $D(S^1_\top) \to F(S^1_\top)$ induced by τ (and similarly for $H_0(\sigma)$ and $H_0(\tau \circledast \sigma)$). This problem can be blamed on the bad functorial properties of L^2_{iso} (used in the definition of Ψ). However, Ω is still natural with respect to natural isomorphisms of defects.

There are two ways of dealing with the above situation: 1. Restrict to the groupoid part of CN_0 and of CN_1; 2. Replace the L^2_{iso} by L^2 in the definition of Ψ—the price to pay for this change is that Ω is then no longer unitary. Both options seem to have shortcomings—in our exposition, we have opted for the first option. One unfortunate consequence of the failure of commutativity of (6.9) is that given defects D, E, F, G as above, and given dualizable sectors ${}_DH_F$ and ${}_EK_G$ with normalized duals[12] $(\bar{H}, r_H, s_H)$ and $(\bar{K}, r_K, s_K)$ the horizontal composition $(\bar{H} \boxtimes_{\mathcal{B}} \bar{K}, r, s)$ of those two normalized duals is not a normalized dual for $H \boxtimes_{\mathcal{B}} K$. (Here, the unit and counit r and s are given by the obvious formulas in terms of r_H, r_K, s_H, and s_K.)

6.B. The $1 \boxtimes 1$-isomorphism for an identity defect

We now show that, if one of the two defects in (6.3) is an identity defect, then the $1 \boxtimes 1$-isomorphism admits a much simpler description, in terms of certain natural transformations

that we describe below. This section and the following Section 6.C are concerned with the behavior of horizontal units in the 3-category of conformal nets; they are more technical and are not needed for the subsequent treatment of the fundamental interchange isomorphism in Section 6.D.

Transformations for fusion of vacuum sectors. Let S_l, S_r, S_b, I, j_l, and j_r be as in Section 4.D, and let $I_l := j_l(I)$ and $I_r := j_r(I)$. We draw them here for convenience:

(6.10)

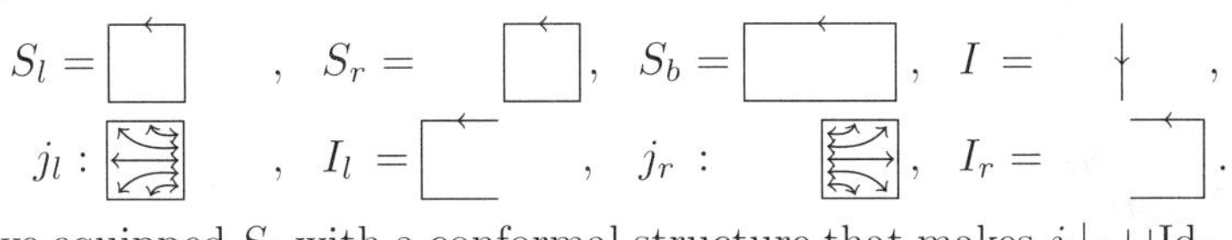

Recall that we equipped S_b with a conformal structure that makes $j_l|_{I_l} \cup \mathrm{Id}_{\mathrm{I}_r} : S_b \to S_r$ and $\mathrm{Id}_{\mathrm{I}_l} \cup j_r|_{I_r} : S_b \to S_l$ conformal (and therefore smooth).

[12]See equations (4.2) and (4.3) in [**BDH11**].

Fix a small number ε. Then Υ^l and Υ^r are invertible natural transformations

$$\mathcal{A}(\partial^{\sqsubset}([1,3/2-\varepsilon]\times[0,1]))\text{-modules} \substack{\rightarrow\\ \Downarrow\\ \rightarrow} \mathcal{A}(\partial^{\sqsubset}([0,3/2-\varepsilon]\times[0,1]))\text{-modules}$$

and

$$\mathcal{A}(\partial^{\sqsupset}([1/2+\varepsilon,1]\times[0,1]))\text{-modules} \substack{\rightarrow\\ \Downarrow\\ \rightarrow} \mathcal{A}(\partial^{\sqsupset}([1/2+\varepsilon,2]\times[0,1]))\text{-modules};$$

that is,

(6.11)
$$\begin{aligned} \Upsilon^l &: \mathcal{A}(\;\;)\text{-modules} \substack{\rightarrow\\ \Downarrow\\ \rightarrow} \mathcal{A}(\;\;)\text{-modules} \\ \Upsilon^r &: \mathcal{A}(\;\;)\text{-modules} \substack{\rightarrow\\ \Downarrow\\ \rightarrow} \mathcal{A}(\;\;)\text{-modules}. \end{aligned}$$

The natural transformation Υ^l goes from the functor $H_0(S_r,\mathcal{A})\boxtimes_{\mathcal{A}(I)} -$ to the functor of restriction along $\mathcal{A}(j_l|_{I_l}\cup \mathrm{Id}) : \mathcal{A}(\quad) \xrightarrow{\cong} \mathcal{A}(\quad)$. Its value on an $\mathcal{A}(\quad)$-module K is given by

$$\Upsilon^l_K : H_0(S_l,\mathcal{A})\boxtimes_{\mathcal{A}(I)} K \xrightarrow{w\otimes 1} L^2(\mathcal{A}(I))\boxtimes_{\mathcal{A}(I)} K \cong K,$$

where w is the isomorphism obtained by composing the canonical identification $v_{I_l} : H_0(S_l,\mathcal{A}) \to L^2(\mathcal{A}(I_l))$ from Appendix C.II and the map $L^2(\mathcal{A}(I_l)) \to L^2(\mathcal{A}(I))$ induced by $j_l : I_l \to I$. Similarly, the natural transformation Υ^r goes from the functor $-\boxtimes_{\mathcal{A}(I)} H_0(S_r,\mathcal{A})$ to the functor of restriction along $\mathcal{A}(\mathrm{Id}\cup j_r|_{I_r}) : \mathcal{A}(\quad) \xrightarrow{\cong} \mathcal{A}(\quad)$. It is given by

$$\Upsilon^r_K : K\boxtimes_{\mathcal{A}(I)} H_0(S_r,\mathcal{A}) \xrightarrow{1\otimes v} K\boxtimes_{\mathcal{A}(I)} L^2(\mathcal{A}(I)) \cong K,$$

where $v = v_I$ is the identification $H_0(S_r,\mathcal{A}) \cong L^2(\mathcal{A}(I))$ from Appendix C.II.

The transformations Υ^l and Υ^r generalize the map $\Upsilon : \blacksquare\blacksquare \to \blacksquare$ from (4.28) and (C.6):

Lemma 6.12. *Let $\epsilon_l := j_l|_I \cup \mathrm{Id}_{I_r} \in \mathrm{Conf}(S_r, S_b)$ and $\epsilon_r := \mathrm{Id}_{I_l}\cup j_r|_I \in \mathrm{Conf}(S_l,S_b)$ be as in Section 4.D. The two maps*

$$H_0(\epsilon_l,\mathcal{A})\circ \Upsilon^l_{H_0(S_r,\mathcal{A})},\quad H_0(\epsilon_r,\mathcal{A})\circ\Upsilon^r_{H_0(S_l,\mathcal{A})} : \blacksquare\blacksquare \xrightarrow{\cong} \blacksquare$$

are equal to each other, and are equal to Υ.

Proof. The equality $\Upsilon = H_0(\epsilon_r)\circ\Upsilon^r_{H_0(S_l)}$ follows from the commutativity of the diagram

$$\begin{array}{ccccc} H_0(S_l)\underset{\mathcal{A}(I)}{\boxtimes} H_0(S_r) & \xrightarrow{v_{I_l}\otimes v_I} & L^2(\mathcal{A}(I_l))\underset{\mathcal{A}(I)}{\boxtimes} L^2(\mathcal{A}(I)) \cong L^2(\mathcal{A}(I_l)) & \xrightarrow{v^*_{I_l}} & H_0(S_b) \\ \| & & \downarrow{\scriptstyle v^*_{I_l}} & & \| \\ H_0(S_l)\underset{\mathcal{A}(I)}{\boxtimes} H_0(S_r) & \xrightarrow{1\otimes v_I} & H_0(S_l)\underset{\mathcal{A}(I)}{\boxtimes} L^2(\mathcal{A}(I)) \cong H_0(S_l) & \xrightarrow{H_0(\epsilon_r)} & H_0(S_b) \end{array}$$

where the top row is Υ and the bottom row is $H_0(\epsilon_r,\mathcal{A})\circ\Upsilon^r_{H_0(S_l)}$. The rightmost square commutes by the naturality of the maps v_I, see (C.3).

To see that $H_0(\epsilon_l)\circ\Upsilon^l_{H_0(S_r)} = H_0(\epsilon_r)\circ\Upsilon^r_{H_0(S_l)}$, one contemplates the diagram

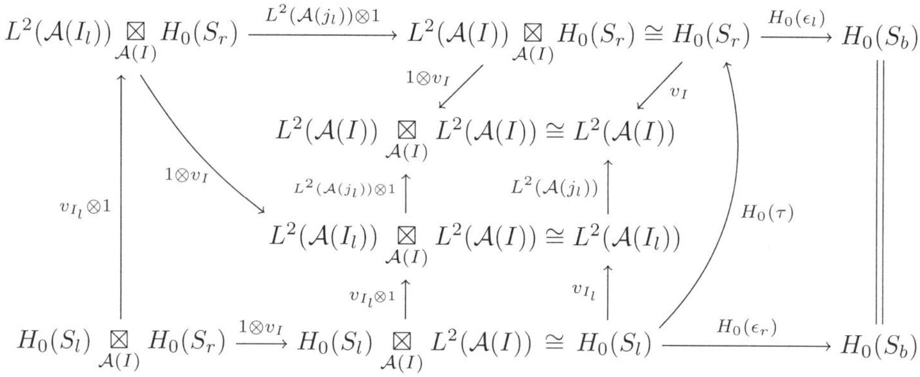

where $\tau := j_l|_{I_l} \cup j_r|_I = \epsilon_l^{-1}\circ\epsilon_r \in \mathrm{Conf}_+(S_l, S_r)$. □

LEMMA 6.13. *The map Υ satisfies the following version of associativity:*

(6.14)

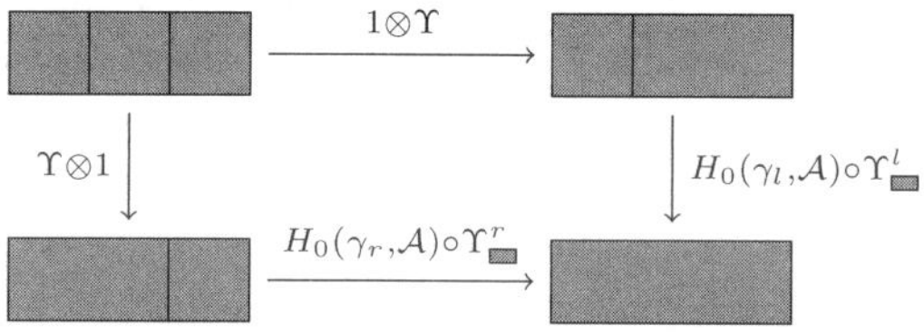

where $\gamma_l : \partial([1,3]\times[0,1]) \to \partial([0,3]\times[0,1])$ and $\gamma_r : \partial([0,2]\times[0,1]) \to \partial([0,3]\times[0,1])$ are the maps

γ_l : 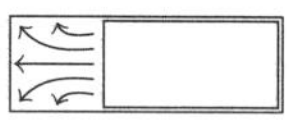 γ_r : 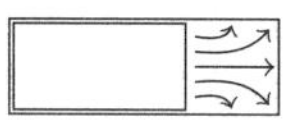

given by $\gamma_l = j_l \cup \mathrm{Id}_{\sqsupset}$ and $\gamma_r = \mathrm{Id}_{\sqsubset} \cup j_r^+$, and j_r^+ is obtained by conjugating j_r by $(x,y)\mapsto(x+1,y)$.

PROOF. Using Lemma 6.12 twice, we can expand (6.14) into the following diagram:

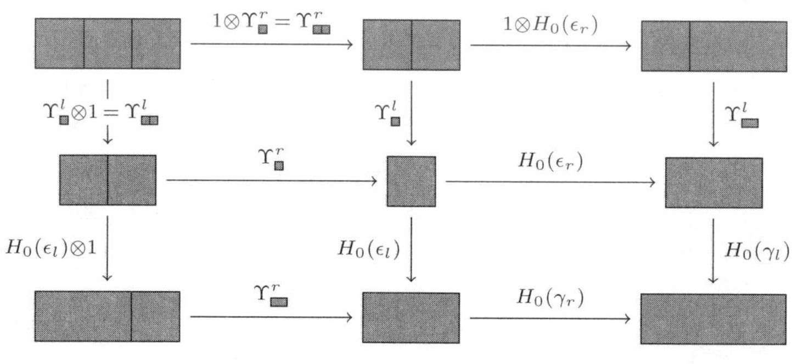

The lower right square commutes by the functoriality of H_0, see (C.1). The remaining three squares commute by the fact that Υ^l and Υ^r are natural transformations. □

The $(1_{\mathrm{id}} \boxtimes 1_D)$-isomorphism as a vacuum fusion transformation. Let ϵ_l and ϵ_r be as above, and let $\epsilon_{l,\top} : S_{r,\top} \to S_{b,\top}$ and $\epsilon_{r,\top} : S_{l,\top} \to S_{b,\top}$ be their restrictions to the upper halves of S_r and S_l, respectively.

LEMMA 6.15. *Let $\mathcal{A}$ be a conformal net with finite index, and let ${}_{\mathcal{A}}D_{\mathcal{B}}$ be an irreducible defect. Let $H_r := H_0(S_r, D)$, where the circle S_r is bicolored as in (4.7). Then the map $\Omega_{\mathrm{id}_{\mathcal{A}},D}$:* 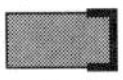 $\to$ 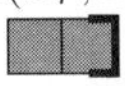 *is the inverse of $L^2(D(\epsilon_{l,\top})) \circ \Upsilon^l_{H_r}$.*

Similarly, assuming instead that $\mathcal{B}$ has finite index, the map $\Omega_{D,\mathrm{id}_{\mathcal{B}}}$:  $\to$  *is the inverse of $L^2(D(\epsilon_{r,\top})) \circ \Upsilon^r_{H_l}$, where $H_l := H_0(S_l, D)$.*

PROOF. We only treat the first equation $\Omega^{-1}_{\mathrm{id}_{\mathcal{A}},D} = L^2(D(\epsilon_{l,\top})) \circ \Upsilon^l_{H_r}$. We first prove it in the case when $D = \mathrm{id}_{\mathcal{A}}$. By definition, $\Omega_{\mathrm{id}_{\mathcal{A}},\mathrm{id}_{\mathcal{A}}}$ is the composite

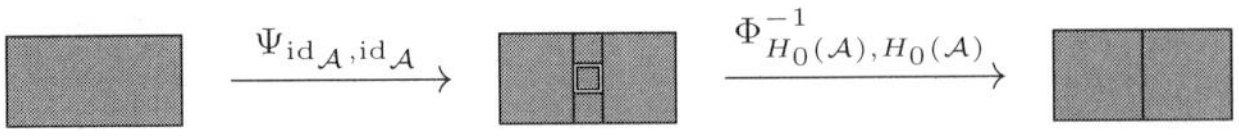

and $\Phi_{H_0(\mathcal{A}),H_0(\mathcal{A})}$ is the composite

It follows that $\Omega_{\mathrm{id}_{\mathcal{A}},\mathrm{id}_{\mathcal{A}}} = \Upsilon^{-1}$ and we are done by Lemma 6.12.

We now treat the general case. By Proposition 4.32 (with the defects $\mathrm{id}_{\mathcal{A}}$, $\mathrm{id}_{\mathcal{A}}$, and D), we have the commutativity of the following diagram:

(6.16)

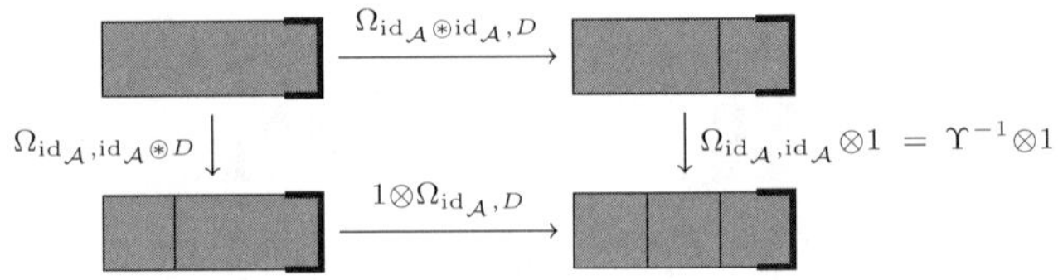

Consider the circles $S_l := \partial([0,1]\times[0,1])$, $S_m := \partial([1,2]\times[0,1])$, $S_r := \partial([2,3]\times[0,1])$, $S_{lm} := \partial([0,2]\times[0,1])$, $S_{mr} := \partial([1,3]\times[0,1])$, $S_{lmr} := \partial([0,3]\times[0,1])$, and the corresponding half-circles $S_{\alpha,\top} := (S_\alpha)_{y\geq\frac{1}{2}}$ for $\alpha \in \{l, m, r, lm, mr, lmr\}$. Let $\varphi : [\frac{1}{2}-\varepsilon, \frac{1}{2}+\varepsilon] \to [\frac{1}{2}-\varepsilon, \frac{3}{2}+\varepsilon]$ and $\psi : [\frac{3}{2}-2\varepsilon, \frac{3}{2}-\varepsilon] \to [\frac{3}{2}-2\varepsilon, \frac{5}{2}-\varepsilon]$ be diffeomorphisms whose derivative is 1 in a neighborhood of the boundary, where ε is a fixed small number. These extend to diffeomorphisms

$$\varphi_{lm} : S_{lm,\top} \to S_{lmr,\top}\,, \qquad \varphi_l : S_{l,\top} \to S_{lm,\top}\,,$$
$$\psi_{lm} : S_{lm,\top} \to S_{lmr,\top}\,, \qquad \psi_m : S_{m,\top} \to S_{mr,\top}\,,$$

whose derivative is 1 outside the domains of φ and ψ, respectively. Let also $\chi := \psi^{-1}_{lm} \circ \varphi_{lm}$. Note that $\chi(x,y) = (x,y)$ for $x \geq \frac{3}{2}$. Since the construction of Ω is natural with respect to isomorphisms, the following squares $\boxed{1}$ are commutative:

(6.17)

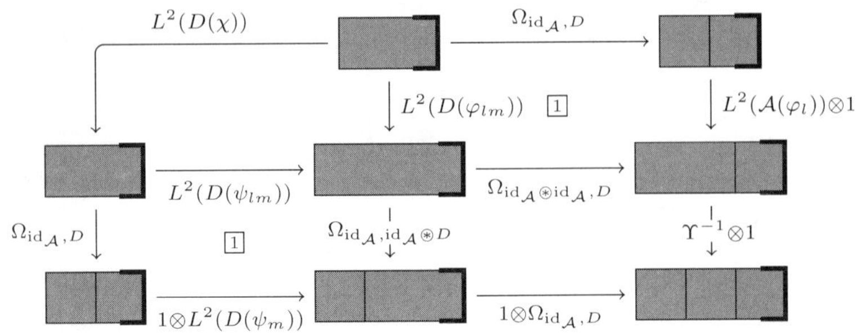

The remaining two squares of (6.17) are commutative by (6.16), and by the definition of χ.

We now consider the following diagram

(6.18)

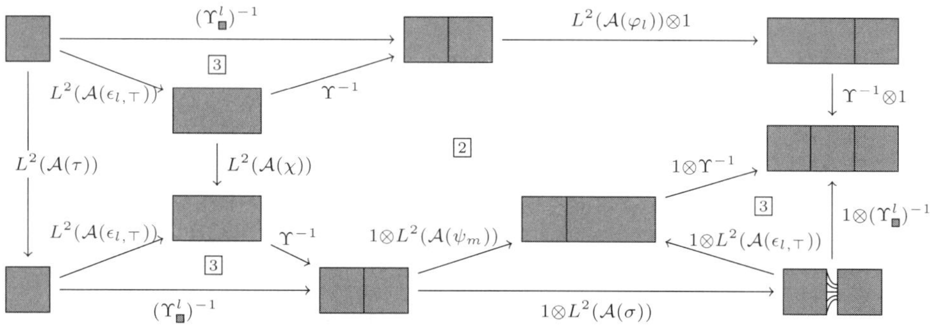

where $\tau = \epsilon_{l,\top}^{-1}\circ\chi\circ\epsilon_{l,\top}$, $\sigma = \epsilon_{l,\top}^{-1}\circ\psi_m$, the lower right corner stands for the fusion of $H_0(S_l)$ with $H_0(S_r)$ along $\mathcal{A}(j_l^+|_{\{1\}\times[0,1]})$, and j_l^+ is obtained by conjugating j_l by $(x,y)\mapsto(x+1,y)$:

j_l^+ : 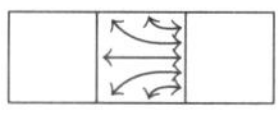 , $j_l^+|_{\{1\}\times[0,1]}$: 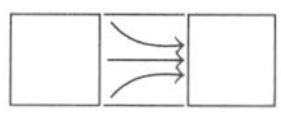 .

Using the identity $\Omega_{\mathrm{id}_\mathcal{A},\mathrm{id}_\mathcal{A}} = \Upsilon^{-1}$ proved earlier, the case $D=\mathrm{id}_\mathcal{A}$ of (6.17) implies the commutativity of 2. The triangles 3 commute by virtue of Lemma 6.12, and so the whole diagram (6.18) is commutative.

Let $\hat\tau,\hat\sigma\in\mathrm{Diff}(\partial[0,1]^2)$ be the symmetric extensions of τ and σ, so that $\hat\tau|_{S^1_\top}=\tau$ and $\hat\sigma|_{S^1_\top}=\sigma$. Note that by definition, both $\hat\tau$ and $\hat\sigma$ commute with $(x,y)\mapsto(x,1-y)$. From the fact that $\chi(x,y)=(x,y)$ for $x\geq\frac{3}{2}$, it follows that $\hat\sigma(x,y)=\hat\tau(x,y)=(x,y)$ for $x\geq\frac{1}{2}$. Let $u\in\mathcal{A}([\ \ )$ and $v\in\mathcal{A}([\ \ )$ be the canonical quantizations, as in the discussion following equation (1.17), of the symmetric diffeomorphisms $\hat\tau$ and $\hat\sigma$. By the definition of these quantizations, we have $L^2(\mathcal{A}(\tau))=\pi(u)$ and $L^2(\mathcal{A}(\sigma))=\pi(v)$, where π is the action of $\mathcal{A}([\ \ )$ on $\blacksquare = H_0(\mathcal{A})$.

We now consider the following diagram of natural transformations between functors from $\mathcal{A}([\ \ )$-modules to Hilbert spaces:

(6.19)

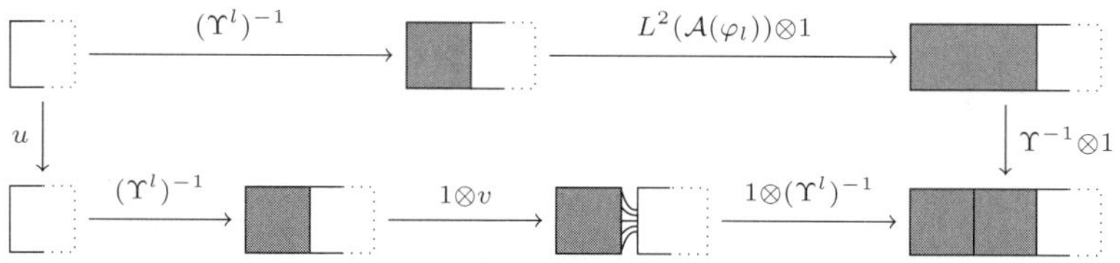

When evaluated on $H_0(\mathcal{A})$, the above diagram commutes by (6.18). Therefore, by Lemma B.24, since $H_0(\mathcal{A})$ is a faithful $\mathcal{A}([\ \ )$-module, the diagram (6.19) commutes regardless of the module one evaluates it on. We now consider the following variant of diagram (6.18):

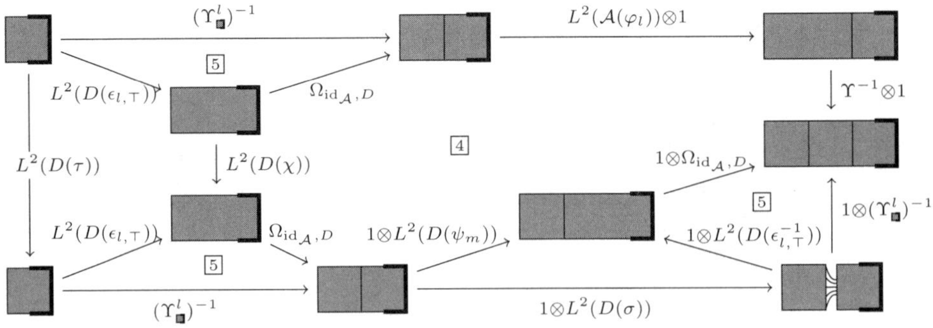

Our goal is to show is that the triangles 5 are commutative. Since D is irreducible, there exists an invertible complex number λ such that

$$\Omega_{\mathrm{id}_{\mathcal{A}},D}\circ L^2(D(\epsilon_{l,\top}))=\lambda\,(\Upsilon^l_{■})^{-1}.$$

The 7-gon 4 is simply (6.17), and it is therefore commutative. The triangle 5 occurs two times with a given orientation, and once with the opposite orientation: the outer 7-gon therefore commutes up to a factor of λ. But using Lemma 1.18 note that outer 7-gon is an instance of (6.19), and is therefore commutative. It follows that $\lambda = 1$. □

6.C. Unitors for horizontal fusion of sectors

In this section, we will introduce variants of the transformations Υ^l and Υ^r that function as unitors for horizontal fusion and that will be more convenient than Υ^l and Υ^r for verifying (in [**BDH16**]) that conformal nets form a 3-category (more precisely, an internal dicategory in the 2-category of symmetric monoidal categories [**DH12**, Def. 3.3]; see Footnote [14]).

We will again be using the circles S_l, S_r, S_b, the intervals I, I_l, I_r, and the involutions $j_l\in\mathrm{Conf}_-(S_l)$, $j_r\in\mathrm{Conf}_-(S_r)$ from (6.10). Let $\alpha_l := j_l|_{I_l}\cup\mathrm{id}_{I_r}\colon S_b\to S_r$ and $\alpha_r := \mathrm{id}_{I_l}\cup j_r|_I\colon S_b\to S_l$ be the diffeomorphisms used in the definition of Υ^l and Υ^r—their inverses appeared in Lemma 6.12 under the names ϵ_l and ϵ_r.

The restriction $\alpha_l|_{\{0\}\times[0,1]}$ is not the map $(0,y)\mapsto(1,y)$ and, as a consequence, the way Υ^l interacts with horizontal fusion is somewhat complicated to describe. (see the lower right corner of (6.18) and the vertical arrow from it); a similar story holds for the restriction of α_r to $\{2\}\times[0,1]$. Our next goal is to introduce diffeomorphisms $\beta_l\colon S_b\to S_r$ and $\beta_r\colon S_b\to S_l$ to replace α_l and α_r, so that the corresponding variants of Υ^l and Υ^r avoid this complication with horizontal fusion.

Pick intervals $I_l^+\subset S_b$ and $I_r^+\subset S_b$ that are sightly larger than I_l and I_r, and diffeomorphisms β_l and β_r that satisfy

$$\beta_l(0,y)=(1,y),\quad \beta_l\circ j=j\circ\beta_l\,,\quad \beta_l|_{S_b\setminus I_l^+}=\mathrm{id}$$
$$\beta_r(2,y)=(1,y),\quad \beta_r\circ j=j\circ\beta_r,\quad \beta_r|_{S_b\setminus I_r^+}=\mathrm{id},$$

where $j(x,y)=(x,1-y)$:

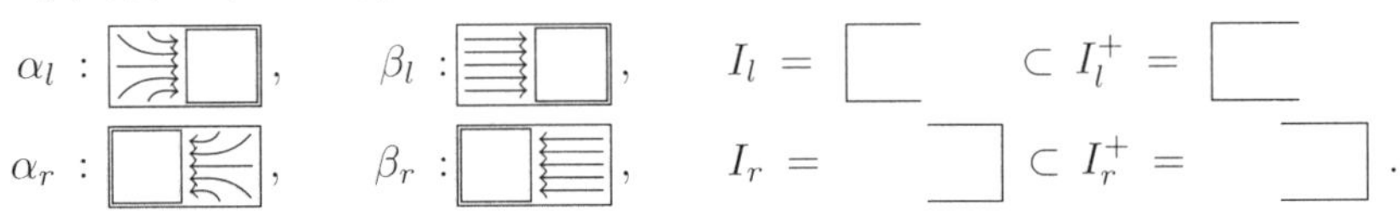

The illustrations of β_l and β_r are somewhat rough, in that they do not record the nontrivial contraction on the horizontal segments of I_l^+ and I_r^+.

The composition $\beta_l \circ \alpha_l^{-1} : S_r \to S_r$ is symmetric with respect to j, and restricts to the identity on the complement of $\alpha_l(I_l^+)$. As explained in the discussion preceding Lemma 1.18, there is a canonical implementation u_l of the diffeomorphism $\beta_l \circ \alpha_l^{-1}$ on $H_0(S_r)$. Similarly, there is a canonical implementation u_r of $\beta_r \circ \alpha_r^{-1}$ on $H_0(S_l)$.

Given a D-E-sector H between $\mathcal{A}$-$\mathcal{B}$-defects D and E, pulling back H along $\mathcal{A}(\alpha_l)$ produces an $(\mathrm{id}_\mathcal{A} \circledast D)$-$(\mathrm{id}_\mathcal{A} \circledast E)$-sector; cf. Lemma 3.1 and the subsequent discussion. This operation is a functor $\alpha_l^* : \mathsf{CN}_2 \to \mathsf{CN}_2$. Similarly, we have functors β_l^*, α_r^*, $\beta_r^* : \mathsf{CN}_2 \to \mathsf{CN}_2$. Multiplication by u_l and u_r then provide natural isomorphisms

$$U^l : \alpha_l^* \cong \beta_l^* : \mathsf{CN}_2 \rightrightarrows \mathsf{CN}_2 , \qquad U^r : \alpha_r^* \cong \beta_r^* : \mathsf{CN}_2 \rightrightarrows \mathsf{CN}_2$$

between the functors α_l^* and β_l^*, and between the functors α_r^* and β_r^*. Recall that Υ^l and Υ^r are natural isomorphisms $H_0(\mathcal{A}) \boxtimes_\mathcal{A} - \cong \alpha_l^*$ and $- \boxtimes_\mathcal{B} H_0(\mathcal{B}) \cong \alpha_r^*$ of functors from CN_2 to CN_2.

Definition 6.20. The left and right *unitors* are the natural transformations $\hat{\Upsilon}^l := U^l \circ \Upsilon^l : H_0(\mathcal{A}) \boxtimes_\mathcal{A} - \to \beta_l^*$ and $\hat{\Upsilon}^r := U^r \circ \Upsilon^r : - \boxtimes_\mathcal{B} H_0(\mathcal{B}) \to \beta_r^*$. We draw them as

$\hat{\Upsilon}^l_H :$ 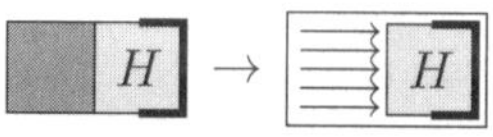

$\hat{\Upsilon}^r_H :$ 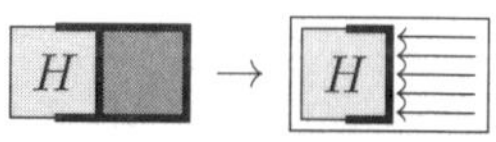

.

(Of course, the precise transformations $\hat{\Upsilon}^l$ and $\hat{\Upsilon}^r$ depend on our unspecified choices of β_l and β_r.)

We record some properties of the unitors that will be used in our later paper [**BDH16**]. There is the following compatibility with the $1 \boxtimes 1$-isomorphism Ω:

Lemma 6.21. *Let $\mathcal{B}$ be a conformal net with finite index, and let ${}_\mathcal{A}D_\mathcal{B}$ and ${}_\mathcal{B}E_\mathcal{C}$ be defects. The unitors commute with the $1 \boxtimes 1$-isomorphism in the sense that*

$$\begin{aligned}
\beta_l^* \Omega_{D,E} \circ \hat{\Upsilon}^l_{H_0(D \circledast_\mathcal{B} E)} &= \hat{\Upsilon}^l_{H_0(D) \boxtimes_\mathcal{B} H_0(E)} \circ \big(1_{H_0(\mathcal{A})} \boxtimes_\mathcal{A} \Omega_{D,E}\big),\\
\beta_r^* \Omega_{D,E} \circ \hat{\Upsilon}^r_{H_0(D \circledast_\mathcal{B} E)} &= \hat{\Upsilon}^r_{H_0(D) \boxtimes_\mathcal{B} H_0(E)} \circ \big(\Omega_{D,E} \boxtimes_\mathcal{C} 1_{H_0(\mathcal{C})}\big).
\end{aligned}$$

That is, the two diagrams

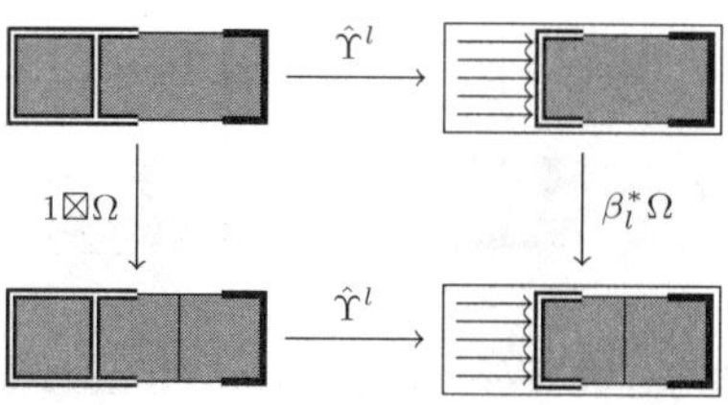

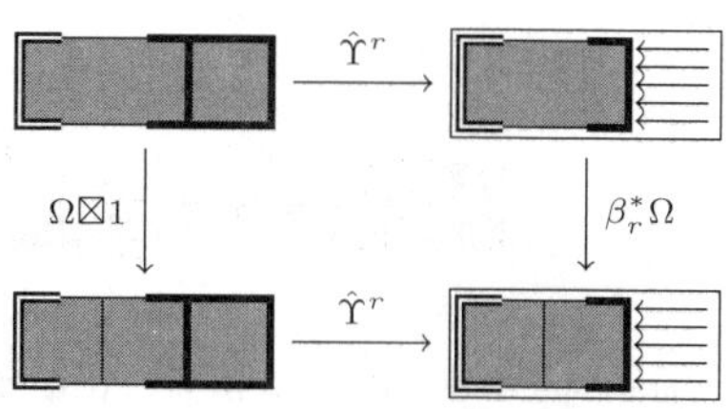

commute.

PROOF. These equalities follow from the naturality of $\hat{\Upsilon}^l$ and $\hat{\Upsilon}^r$, specifically by viewing them as a natural transformations

$\hat{\Upsilon}^l$: $\mathcal{A}($$)$-modules $\Rrightarrow$ $\mathcal{A}($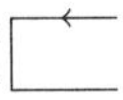$)$-modules

$\hat{\Upsilon}^r$: $\mathcal{C}($$)$-modules $\Rrightarrow$ $\mathcal{C}($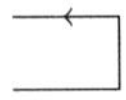$)$-modules. □

Let $\alpha_{l,\top}, \beta_{l,\top} : S_{b,\top} \to S_{r,\top}$ and $\alpha_{r,\top}, \beta_{r,\top} : S_{b,\top} \to S_{l,\top}$ be the restrictions of α_l, β_l, α_r, β_r to the upper half of S_b.

LEMMA 6.22. *Let ${}_{\mathcal{A}}D_{\mathcal{B}}$ be a defect, color the circles S_r and S_l as in (4.7), and let $H_r := H_0(S_r, D)$ and $H_l := H_0(S_l, D)$.*

Then the map $\hat{\Upsilon}^l_{H_r} \circ \Omega_{\mathrm{id}_{\mathcal{A}},D}$: 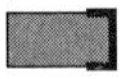 $\to$ 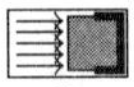 *coincides with $L^2(D(\beta_{l,\top}))$, and $\hat{\Upsilon}^r_{H_l} \circ \Omega_{D,\mathrm{id}_{\mathcal{B}}}$:*  $\to$ 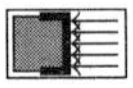 *coincides with $L^2(D(\beta_{r,\top}))$.*

PROOF. From Lemma 6.15 (recall that $\epsilon_l = \alpha_l^{-1}$ and $\epsilon_r = \alpha_r^{-1}$), we know that

$$\Upsilon^l_{H_r} \circ \Omega_{\mathrm{id}_{\mathcal{A}},D} = L^2(D(\alpha_{l,\top}))\ , \qquad \Upsilon^r_{H_l} \circ \Omega_{D,\mathrm{id}_{\mathcal{B}}} = L^2(D(\alpha_{r,\top})).$$

We also have $U^l_{H_r} = L^2(D(\beta_{l,\top} \circ \alpha_{l,\top}^{-1}))$ and $U^r_{H_l} = L^2(D(\beta_{r,\top} \circ \alpha_{r,\top}^{-1}))$ by Lemma 1.18. Finally recall that $\hat{\Upsilon}^l = U^l \circ \Upsilon^l$ and $\hat{\Upsilon}^r = U^r \circ \Upsilon^r$. We therefore have

$$\begin{aligned}\hat{\Upsilon}^l_{H_r} \circ \Omega_{\mathrm{id}_{\mathcal{A}},D} &= U^l_{H_r} \circ \Upsilon^l_{H_r} \circ \Omega_{\mathrm{id}_{\mathcal{A}},D} \\ &= L^2(D(\beta_{l,\top} \circ \alpha_{l,\top}^{-1})) \circ L^2(D(\alpha_{l,\top})) = L^2(D(\beta_{l,\top}))\end{aligned}$$

and

$$\begin{aligned}\hat{\Upsilon}^r_{H_l} \circ \Omega_{D,\mathrm{id}_{\mathcal{B}}} &= U^r_{H_l} \circ \Upsilon^r_{H_l} \circ \Omega_{D,\mathrm{id}_{\mathcal{B}}} \\ &= L^2(D(\beta_{r,\top} \circ \alpha_{r,\top}^{-1})) \circ L^2(D(\alpha_{r,\top})) = L^2(D(\beta_{r,\top})).\end{aligned}$$ □

Combining this lemma with the factorization of $L^2(D(\beta_{l,\top}))$ and $L^2(D(\beta_{r,\top}))$ from (B.7), we obtain the commutativity of the following two diagrams:

(6.23)
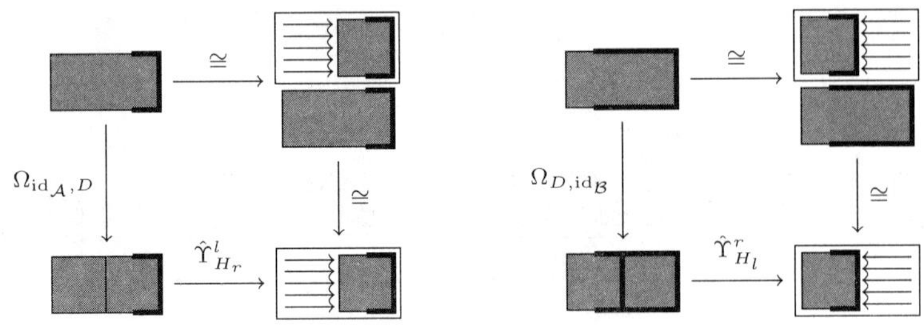

Here, the stacked pictures in the upper right-hand corners signify the Connes fusion along the algebras D(——) and D(——) respectively, and we have suppressed the isomorphism (B.6).

6.D. The interchange isomorphism

In a 2-category, the *interchange law* says that the two ways of evaluating the diagram

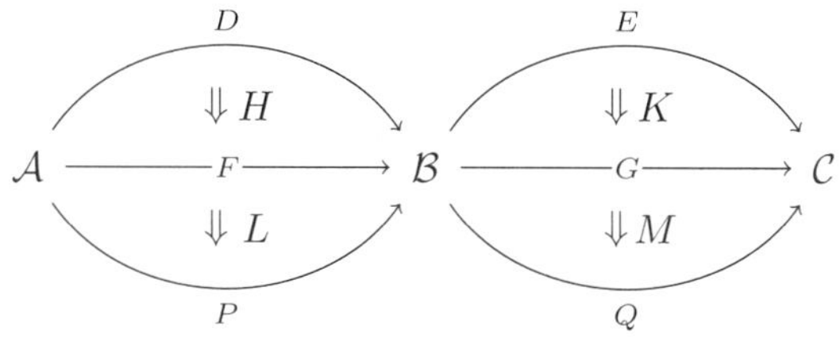

are equal to each other: if one first performs the two vertical compositions and then composes horizontally, or one first composes horizontally and then vertically, one should obtain the same result. In our case, in a 3-category, the two ways of fusing four sectors

(6.24) 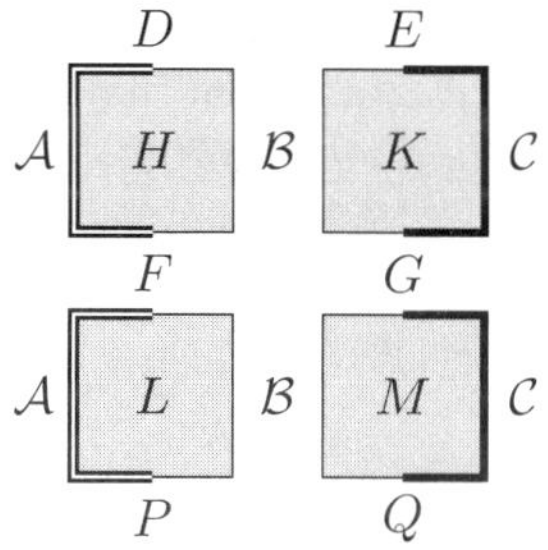

,

namely 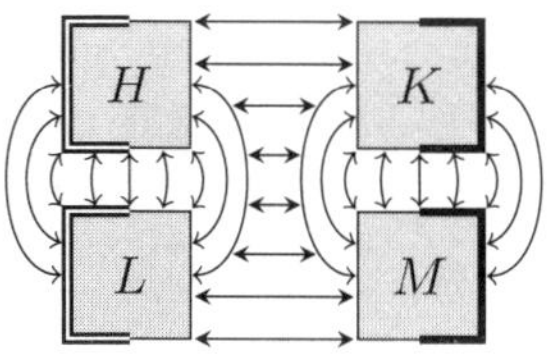

versus 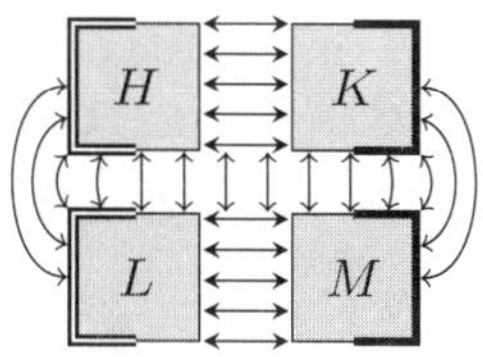

,

should yield the same answer up to natural isomorphism. In other words, we need a unitary natural transformation[13]

$$(6.25)\qquad \left(\mathsf{CN}_2 \times_{\mathsf{CN}_1^f} \mathsf{CN}_2\right) \times_{\mathsf{CN}_0^f} \left(\mathsf{CN}_2 \times_{\mathsf{CN}_1^f} \mathsf{CN}_2\right) \overset{\Downarrow}{\rightrightarrows} \mathsf{CN}_2$$

between the functors $\mathsf{fusion}_\mathsf{h} \circ (\mathsf{fusion}_\mathsf{v} \times \mathsf{fusion}_\mathsf{v})$ and $\mathsf{fusion}_\mathsf{v} \circ (\mathsf{fusion}_\mathsf{h} \times \mathsf{fusion}_\mathsf{h}) \circ \tau$, where

$$\begin{aligned}\tau\ :\ &\left(\mathsf{CN}_2 \times_{\mathsf{CN}_1^f} \mathsf{CN}_2\right) \times_{\mathsf{CN}_0^f} \left(\mathsf{CN}_2 \times_{\mathsf{CN}_1^f} \mathsf{CN}_2\right)\\ &\to \left(\mathsf{CN}_2 \times_{\mathsf{CN}_0^f} \mathsf{CN}_2\right) \times_{\mathsf{CN}_1^f \times_{\mathsf{CN}_0^f} \mathsf{CN}_1^f} \left(\mathsf{CN}_2 \times_{\mathsf{CN}_0^f} \mathsf{CN}_2\right)\end{aligned}$$

is the isomorphism that exchanges the two middle factors.

More concretely, given sectors ${}_DH_F$, ${}_EK_G$, ${}_FL_P$, ${}_GM_Q$ as in (6.24), we are looking for a unitary isomorphism

$$(6.26)\qquad \big(H \boxtimes_F L\big) \boxtimes_{\mathcal{B}} \big(K \boxtimes_G M\big) \xrightarrow{\cong} \big(H \boxtimes_{\mathcal{B}} K\big) \boxtimes_{F \circledast_{\mathcal{B}} G} \big(L \boxtimes_{\mathcal{B}} M\big)$$

of $D \circledast_{\mathcal{B}} E$ - $P \circledast_{\mathcal{B}} Q$ -sectors.

---

[13]Here, as in the $1 \boxtimes 1$-isomorphism Ω, we restrict to the groupoid parts of CN_1 and CN_0.

We may view (H, K, L, M) as an object of the category

$$\mathsf{C} := \begin{array}{l} F\big(\sqcup\big)\text{-modules} \times G\big(\sqcup\big)\text{-modules} \\ \times\; F\big(\sqcap\big)\text{-modules} \times G\big(\sqcap\big)\text{-modules} \end{array}.$$

The forgetful functor $(\mathsf{CN}_2 \times_{\{F\}} \mathsf{CN}_2) \times (\mathsf{CN}_2 \times_{\{G\}} \mathsf{CN}_2) \to \mathsf{C}$ is faithful. In order to construct the natural transformation (6.25), it is therefore enough to produce corresponding natural transformations

$$\mathsf{C} \;\overset{\Downarrow}{\rightrightarrows}\; \text{Hilbert Spaces} \tag{6.27}$$

for every F and G. The fact that (6.26) intertwines the actions of $D(\ulcorner)$, $E(\urcorner)$ $P(\llcorner)$, and $Q(\lrcorner)$, i.e., that it is a morphism of $D \circledast_{\mathcal{B}} E$-$P \circledast_{\mathcal{B}} Q$-sectors, will then follow from the naturality of (6.27).

The object $\big(H_0(F), H_0(G), H_0(F), H_0(G)\big) \in \mathsf{C}$ consists of faithful modules; the obvious analog of Lemma 4.3 (itself a generalization of B.24) applies, and so it is enough to construct the natural transformation (6.27) on this object. Using the isomorphisms (6.3) and (2.19), the required transformation is given by

$$\begin{aligned} &\big(H_0(F) \boxtimes_F H_0(F)\big) \boxtimes_{\mathcal{B}} \big(H_0(G) \boxtimes_G H_0(G)\big) \xrightarrow{\cong} H_0(F) \boxtimes_{\mathcal{B}} H_0(G) \\ &\xrightarrow{\Omega^{-1}} H_0\big(F \circledast_{\mathcal{B}} G\big) \xrightarrow{\cong} H_0\big(F \circledast_{\mathcal{B}} G\big) \boxtimes_{F \circledast_{\mathcal{B}} G} H_0\big(F \circledast_{\mathcal{B}} G\big) \\ &\xrightarrow{\Omega \boxtimes \Omega} \big(H_0(F) \boxtimes_{\mathcal{B}} H_0(G)\big) \boxtimes_{F \circledast_{\mathcal{B}} G} \big(H_0(F) \boxtimes_{\mathcal{B}} H_0(G)\big). \end{aligned}$$

Using the compatibility of Ω with the monoidal structure (Proposition 6.6), the same can be deduced for the interchange isomorphism: (6.25) is a monoidal natural transformation.

APPENDIX A

Components for the 3-category of conformal nets

The purpose of our next paper [**BDH16**] is to construct the symmetric monoidal 3-category of conformal nets. More precisely, we will construct an internal dicategory object $(\mathsf{C}_0, \mathsf{C}_1, \mathsf{C}_2)$ in the 2-category of symmetric monoidal categories [**DH12**, Def. 3.3].[14] In this book, we have developed the essential ingredients of that 3-category. These ingredients are:

- A symmetric monoidal category C_0 whose objects are the conformal nets with finite index, and whose morphisms are the isomorphisms between them.
- A symmetric monoidal category C_1 whose objects are the defects between conformal nets of finite index, and whose morphisms are the isomorphisms.
- A symmetric monoidal category C_2 whose objects are sectors (between defects between conformal nets of finite index), and whose morphisms are those homomorphisms of sectors that cover isomorphisms of defects and of conformal nets.
- These come with source and target functors $\mathsf{s}, \mathsf{t} : \mathsf{C}_1 \to \mathsf{C}_0$ and $\mathsf{s}, \mathsf{t} : \mathsf{C}_2 \to \mathsf{C}_1$ subject to the identities $\mathsf{s} \circ \mathsf{s} = \mathsf{s} \circ \mathsf{t}$ and $\mathsf{t} \circ \mathsf{s} = \mathsf{t} \circ \mathsf{t}$.
- A symmetric monoidal functor $\mathsf{composition} : \mathsf{C}_1 \times_{\mathsf{C}_0} \mathsf{C}_1 \to \mathsf{C}_1$ that describes the composition (or fusion) of defects (1.48). That the composition of defects exists is the content of Theorem 1.44.
- A symmetric monoidal functor $\mathsf{fusion}_\mathsf{h} : \mathsf{C}_2 \times_{\mathsf{C}_0} \mathsf{C}_2 \to \mathsf{C}_2$ providing the horizontal composition of sectors (2.12).
- A symmetric monoidal functor $\mathsf{fusion}_\mathsf{v} : \mathsf{C}_2 \times_{\mathsf{C}_1} \mathsf{C}_2 \to \mathsf{C}_2$ providing the vertical composition of sectors (2.16).
- Symmetric monoidal functors $\mathsf{identity} : \mathsf{C}_0 \to \mathsf{C}_1$ and $\mathsf{identity}_\mathsf{v} : \mathsf{C}_1 \to \mathsf{C}_2$ providing the identity defects (1.39) and identity sectors (2.8).
- A monoidal natural transformation $\mathsf{associator}$: $\mathsf{C}_1 \times_{\mathsf{C}_0} \mathsf{C}_1 \times_{\mathsf{C}_0} \mathsf{C}_1 \overset{\Downarrow}{\rightrightarrows} \mathsf{C}_1$ that is an associator for $\mathsf{composition}$ (1.57).
- A monoidal natural transformation $\mathsf{associator}_\mathsf{h} : \mathsf{C}_2 \times_{\mathsf{C}_0} \mathsf{C}_2 \times_{\mathsf{C}_0} \mathsf{C}_2 \overset{\Downarrow}{\rightrightarrows} \mathsf{C}_2$ that is an associator for $\mathsf{fusion}_\mathsf{h}$ (2.13)
- A monoidal natural transformation $\mathsf{associator}_\mathsf{v} : \mathsf{C}_2 \times_{\mathsf{C}_1} \mathsf{C}_2 \times_{\mathsf{C}_1} \mathsf{C}_2 \overset{\Downarrow}{\rightrightarrows} \mathsf{C}_2$ that is an associator for $\mathsf{fusion}_\mathsf{v}$ (2.17).
- Two monoidal natural transformations $\mathsf{unitor}_\mathsf{t}, \mathsf{unitor}_\mathsf{b} : \mathsf{C}_2 \overset{\Downarrow}{\rightrightarrows} \mathsf{C}_2$ that relate $\mathsf{fusion}_\mathsf{v}$ and $\mathsf{identity}_\mathsf{v}$ (2.18).
- The coherences for $\mathsf{composition}$ and $\mathsf{identity}$ are "weak": instead of natural transformations $\mathsf{C}_1 \overset{\Downarrow}{\rightrightarrows} \mathsf{C}_1$, we have four functors $\mathsf{unitor}_\mathsf{tl}, \mathsf{unitor}_\mathsf{tr}, \mathsf{unitor}_\mathsf{bl}, \mathsf{unitor}_\mathsf{br} : \mathsf{C}_1 \to \mathsf{C}_2$ (3.2, 3.3). This weakness, which is an intrinsic feature of

[14] Following [**DH12**], a dicategory is a bicategory where the associators $(fg)h \Rightarrow f(gh)$ are strict (i.e., they are identity 2-morphisms) but the unitors $1f \Rightarrow f$ and $f1 \Rightarrow f$ are not necessarily strict.

conformal nets and defects, is what forces us to use the notion of internal dicategory [**DH12**, Def. 3.3] instead of the simpler notion of internal 2-category [**DH12**, Def. 3.1].

- The coherence between composition and $\mathsf{identity}_\mathsf{v}$ is a monoidal natural transformation $\mathsf{C}_1 \times_{\mathsf{C}_0} \mathsf{C}_1 \circlearrowleft \mathsf{C}_2$ (6.1, 6.3). The difficult construction of this coherence forces us to restrict the morphisms in the category C_1 of defects to be isomorphisms. We have called this natural transformation the "$1 \boxtimes 1$-isomorphism" because its domain is a Connes fusion of two identity sectors.
- Finally, the fundamental interchange isomorphism, a coherence between $\mathsf{fusion}_\mathsf{h}$ and $\mathsf{fusion}_\mathsf{v}$, is a monoidal natural transformation

$$(\mathsf{C}_2 \times_{\mathsf{C}_1} \mathsf{C}_2) \times_{\mathsf{C}_0} (\mathsf{C}_2 \times_{\mathsf{C}_1} \mathsf{C}_2) \circlearrowleft \mathsf{C}_2 \quad (6.25).$$

  Its definition relies crucially on the $1 \boxtimes 1$-isomorphism.

APPENDIX B

Von Neumann algebras

Given a Hilbert space H, we let $\mathbf{B}(H)$ denote its algebra of bounded operators. The ultraweak topology on $\mathbf{B}(H)$ is the topology of pointwise convergence with respect to the pairing with its predual, the trace class operators.

DEFINITION B.1. A von Neumann algebra, is a topological *-algebra (without any compatibility between the topology and the algebra structure) that is embeddable as closed subalgebra of $\mathbf{B}(H)$ with respect to the ultraweak topology.

The spatial tensor product $A_1\bar{\otimes}A_2$ of von Neumann algebras $A_i \subset \mathbf{B}(H_i)$ is the closure in $\mathbf{B}(H_1 \otimes H_2)$ of their algebraic tensor product $A_1 \otimes_{alg} A_2$.

DEFINITION B.2. Let A be a von Neumann algebra. A left (right) A-module is a Hilbert space H equipped with a continuous homomorphism from A (respectively A^{op}) to $\mathbf{B}(H)$. We will use the notation ${}_AH$ (respectively H_A) to denote the fact that H is a left (right) A-module.

We now concisely recapitulate those aspects of [**BDH11, BDH13, BDH14**] that are used in the present book, along with some other general facts about von Neumann algebras. For further details, we refer the reader to [**BDH11**, §2 and §6] for Section B.I, to [**BDH11**, §3] for Section B.II, to [**BDH14**, Appendix A] for Section B.III, to [**BDH13**, §1.3] for Section B.IV, to [**BDH11**, §4] for Section B.VI, and to [**BDH11**, §5] for Section B.VII.

B.I. The Haagerup L^2-space

A faithful left module H for a von Neumann algebra A is called a *standard form* if it comes equipped with an antilinear isometric involution J and a selfdual cone $P \subset H$ subject to the properties

(i) $JAJ = A'$ on H,
(ii) $JcJ = c^*$ for all $c \in Z(A)$,
(iii) $J\xi = \xi$ for all $\xi \in P$,
(iv) $aJaJ(P) \subseteq P$ for all $a \in A$

where A' denotes the commutant of A. The operator J is called the *modular conjugation*. The standard form is an A–A-bimodule, with right action $\xi a := Ja^*J\xi$. It is unique up to unique unitary isomorphism [**Haa75**].

The space of continuous linear functionals $A \to \mathbb{C}$ forms a Banach space $A_* = L^1(A)$ called the predual of A. It comes with a positive cone $L^1_+(A) := \{\phi \in A_* \mid \phi(x) \geq 0\ \forall x \in A_+\}$ and two commuting A-actions given by $(a\phi b)(x) := \phi(bxa)$. Given a von Neumann algebra A there is a canonical construction of a standard form for A [**Kos80**]. It is the completion of

$$\bigoplus_{\phi \in L^1_+(A)} \mathbb{C}\sqrt{\phi}$$

with respect to some pre-inner product, and is denoted $L^2(A)$. The positive cone in L^2A is given by $L^2_+(A) := \{\sqrt{\phi} \mid \phi \in L^1_+(A)\}$. The modular conjugation J_A sends $\lambda\sqrt{\phi}$ to $\bar{\lambda}\sqrt{\phi}$ for $\lambda \in \mathbb{C}$.

If $f\colon A \to B$ is an isomorphism, then we write $L^2(f)\colon L^2(A) \xrightarrow{\cong} L^2(B)$ for the induced unitary isomorphism. The standard form is in fact functorial with respect to a bigger class of maps; see Section B.VI.

B.II. Connes fusion

DEFINITION B.3. Given two modules H_A and ${}_AK$ over a von Neumann algebra A, their Connes fusion $H \boxtimes_A K$ is the completion of

$$\operatorname{Hom}\big(L^2(A)_A, H_A\big) \otimes_A L^2(A) \otimes_A \operatorname{Hom}\big({}_AL^2(A), {}_AK\big) \tag{B.4}$$

with respect to the inner product $\big\langle \phi_1 \otimes \xi_1 \otimes \psi_1,\, \phi_2 \otimes \xi_2 \otimes \psi_2 \big\rangle := \big\langle (\phi_2^*\phi_1)\xi_1(\psi_1\psi_2^*), \xi_2 \big\rangle$ [**Con90, Sau83, Was98**]. Here, we have written the action of ψ_i on the right, which means that $\psi_1\psi_2^*$ stands for the composite $L^2(A) \xrightarrow{\psi_1} K \xrightarrow{\psi_2^*} L^2(A)$.

The L^2 space is a unit for Connes fusion in the sense that there are canonical unitary isomorphisms

$${}_AL^2(A) \boxtimes_A H \cong {}_AH \qquad \text{and} \qquad H \boxtimes_A L^2(A)_A \cong H_A \tag{B.5}$$

defined by $\phi \otimes \xi \otimes \psi \mapsto (\phi\xi)\psi$ and $\phi \otimes \xi \otimes \psi \mapsto \phi(\xi\psi)$. If $f\colon A \to B$ is an isomorphism of von Neumann algebras, H_A and ${}_BK$ are modules, then

$$(H_A)_{f^{-1}} \boxtimes_B {}_BK \cong H_A \boxtimes_A {}_f({}_BK) \tag{B.6}$$

via $\phi \otimes \xi \otimes \psi \mapsto (\phi \circ L^2(f)) \otimes L^2(f)^{-1}(\xi) \otimes (\psi \circ L^2(f))$. Here the indices ${}_{f^{-1}}$ and ${}_f$ indicate restrictions of actions along the isomorphisms f and f^{-1}. Using (B.5) $L^2(f)$ can be expressed as

$$L^2(A) \cong L^2(A)_{f^{-1}} \boxtimes_B L^2(B) \cong L^2(A) \boxtimes_A {}_fL^2(B) \cong L^2(B). \tag{B.7}$$

B.III. Cyclic fusion

Let $n \geq 2$ be some number. For $i \in \{1, \ldots, n\}$, let A_i be a von Neumann algebras, and let H_i be $A_i \bar{\otimes} A_{i+1}^{op}$-modules (cyclic numbering). Then for each $i, j \in \{1, \ldots, n\}$, we can form the fusion of $H_i \boxtimes_{A_{i+1}} \ldots \boxtimes_{A_{j-1}} H_{j-1}$ (cyclic numbering) with $H_j \boxtimes_{A_{j+1}} \ldots \boxtimes_{A_{i-1}} H_{i-1}$ over the algebra $A_i^{op} \bar{\otimes} A_j$. The Hilbert space

$$\big(H_i \boxtimes_{A_{i+1}} \ldots \boxtimes_{A_{j-1}} H_{j-1}\big) \underset{A_i^{op} \bar{\otimes} A_j}{\boxtimes} \big(H_j \boxtimes_{A_{j+1}} \ldots \boxtimes_{A_{i-1}} H_{i-1}\big)$$

is independent, up to canonical unitary isomorphism, of the choices of i and j [**BDH14**, Appendix A]. We call the above Hilbert space the cyclic fusion of the H_i's, and denote it by

$$\underbrace{H_1 \boxtimes_{A_2} \cdots \boxtimes_{A_n} H_n \boxtimes_{A_1}}_{\text{cyclic}} .$$

B.IV. Fusion and fiber product of von Neumann algebras

DEFINITION B.8. Let $A \leftarrow C^{op}$, $C \rightarrow B$ be two homomorphisms between von Neumann algebras, and let ${}_AH$ and ${}_BK$ be faithful modules. Viewing H as a right C-module, we may form the Connes fusion $H \boxtimes_C K$. One then defines the fusion of A and B over C as

$$A \circledast_C B := (A \cap C^{op\prime}) \vee (C' \cap B) \subset \mathbf{B}(H \boxtimes_C K), \tag{B.9}$$

where the commutants of C^{op} and C are taken in H and K, respectively.

The fusion is independent, up to canonical isomorphism, of the choice of modules H and K [**BDH13**, Prop. 1.23]. If those modules are not faithful, then there is still an action, albeit non-faithful, of $A \circledast_C B$ on $H \boxtimes_C K$ [**BDH13**, Lem. 1.24]. Note that the operation $\circledast$ is not associative [**BDH13**, Warn. 1.22].

The fiber product of von Neumann algebras (introduced in [**Sau85**] when C is abelian, and in [**Tim08**, Def. 10.2.4] in general) is better behaved:

DEFINITION B.10. In the situation of Definition B.8, the fiber product of A and B over C is given by

$$A *_C B := (A' \otimes_{alg} B')',$$

where the commutants A' and B' are taken in $\mathbf{B}(H)$ and $\mathbf{B}(K)$ respectively, while the last one is taken in $\mathbf{B}(H \boxtimes_C K)$.

The fiber product is independent of the choice of modules H and K and there is an associator $A *_C (B *_D E) \rightarrow (A *_C B) *_D E$ that satisfies the pentagon identity [**Tim08**, Prop. 10.2.8].

If $C = \mathbb{C}$, then $A *_{\mathbb{C}} B = A \circledast_{\mathbb{C}} B$ is the spatial tensor product $A \bar{\otimes} B$ of von Neumann algebras.

B.V. Compatibility with tensor products

There is a canonical isomorphism [**MT84, SW82**]

$$L^2(A) \otimes L^2(B) \cong L^2(A \bar{\otimes} B)$$

that sends $\sqrt{\phi} \otimes \sqrt{\psi}$ to $\sqrt{\phi \otimes \psi}$. This isomorphism provides a natural compatibility between Connes fusion and tensor products,

$$(H_1 \boxtimes_A H_2) \otimes (K_1 \boxtimes_B K_2) \cong (H_1 \otimes K_1) \boxtimes_{A \bar{\otimes} B} (H_2 \otimes K_2).$$

This isomorphism can then be used to construct natural compatibility isomorphisms between the spatial tensor product and the fusion, respectively the fiber product, of von Neumann algebras:

$$\begin{aligned} (A_1 \circledast_{C_1} B_1) \bar{\otimes} (A_2 \circledast_{C_2} B_2) &\cong (A_1 \bar{\otimes} A_2) \circledast_{C_1 \bar{\otimes} C_2} (B_1 \bar{\otimes} B_2), \\ (A_1 *_{C_1} B_1) \bar{\otimes} (A_2 *_{C_2} B_2) &\cong (A_1 \bar{\otimes} A_2) *_{C_1 \bar{\otimes} C_2} (B_1 \bar{\otimes} B_2). \end{aligned}$$

Here, those isomorphisms also rely on the equation $(A \bar{\otimes} B)' = A' \bar{\otimes} B'$ (Tomita's commutator theorem [**Tak70**, Thm. 12.3]).

B.VI. Dualizability

A von Neumann algebra whose center is $\mathbb{C}$ is called a *factor*. Von Neumann algebras with finite-dimensional center are finite direct sums of factors.

DEFINITION B.11. Let A and B be von Neumann algebras with finite-dimensional center. Given an A–B-bimodule H, we say that a B–A-bimodule $\bar{H}$ is dual to H if it comes equipped with maps

$$R\colon {}_AL^2(A)_A \to {}_AH \boxtimes_B \bar{H}_A \qquad S\colon {}_BL^2(B)_B \to {}_B\bar{H} \boxtimes_A H_B \tag{B.12}$$

subject to the duality equations $(R^* \otimes 1)(1 \otimes S) = 1$, $(S^* \otimes 1)(1 \otimes R) = 1$, and to the normalization $R^*(x \otimes 1)R = S^*(1 \otimes x)S$ for all $x \in \mathrm{End}({}_AH_B)$. A bimodule whose dual module exists is called *dualizable*.

If ${}_AH_B$ is a dualizable bimodule, then its dual bimodule is well defined up to canonical unitary isomorphism [**BDH11**, Thm. 4.22]. Moreover, the dual bimodule is canonically isomorphic to the complex conjugate Hilbert space $\overline{H}$, with the actions $b\bar{\xi}a := \overline{a^*\xi b^*}$ [**BDH11**, Cor. 6.12].

A homomorphism $f\colon A \to B$ between von Neumann algebras with finite-dimensional center is said to be *finite* if the associated bimodule ${}_AL^2(B)_B$ is dualizable. If $f\colon A \to B$ is a finite homomorphism, then there is an induced map $L^2(f)\colon L^2(A) \to L^2(B)$, and we have $L^2(f \circ g) = L^2(f) \circ L^2(g)$. In other words, Haagerup's L^2-space is functorial with respect to finite homomorphisms [**BDH11**]. The map $L^2(f)$ is bounded and A-A-bilinear, but usually not isometric.

B.VII. Statistical dimension and minimal index

The statistical dimension of a dualizable bimodule ${}_AH_B$ between factors is given by

$$\dim({}_AH_B) := R^*R = S^*S \in \mathbb{R}_{\geq 0}$$

where R and S are as in (B.12). For non-dualizable bimodules, one declares $\dim({}_AH_B)$ to be ∞. If $A = \oplus A_i$ and $B = \oplus B_j$ are finite direct sums of factors, then we can decompose $H = \oplus H_{ij}$ as a direct sum of A_i–B_j-bimodules and define the matrix-valued statistical dimension

$$\dim({}_AH_B)_{ij} := \dim({}_{A_i}H_{ij\,B_j}).$$

This matrix-valued dimension is additive with respect to addition of modules and multiplicative with respect to Connes fusion [**BDH11**, §5]:

$$\dim({}_AH_B \oplus {}_AK_B) = \dim({}_AH_B) + \dim({}_AK_B) \tag{B.13}$$

$$\dim({}_AH_B \boxtimes_B {}_BK_C) = \dim({}_AH_B) \cdot \dim({}_BK_C). \tag{B.14}$$

Given a finite homomorphism $f : A \to B$ between von Neumann algebras with finite-dimensional center, we let

$$[\![B : A]\!] := \dim({}_AL^2B_B)$$

denote the matrix of statistical dimensions of ${}_AL^2B_B$. If ${}_AH_B$ is a bimodule where B acts faithfully, then by [**BDH11**, Lem. 5.16]

$$\dim({}_AH_B) = [\![B' : A]\!] \tag{B.15}$$

where B' is the commutant of B on H. If A also acts faithfully, then

$$[\![B : A]\!] = [\![A' : B']\!]^T, \tag{B.16}$$

where ^{T} denotes the transposed matrix [**BDH11**, Cor. 5.17]. The minimal index $[B : A]$ of an inclusion of factors $\iota\colon A \to B$ is the square of the statistical dimension of ${}_AL^2B_B$ [**BDH11**, Def. 5.10]. For inclusions $A \subseteq B \subseteq C$ of von Neumann algebras with finite-dimensional center we have

$$[\![C : A]\!] = [\![B : A]\!] \cdot [\![C : B]\!], \tag{B.17}$$

by [**BDH11**, Eq. (5.14)]. Moreover, the inclusion map $A \to B$ is an isomorphism if and only if $[\![B : A]\!]$ is a permutation matrix [**BDH11**, §5]. If C is a factor, then

$$[\![B \bar{\otimes} C : A \bar{\otimes} C]\!] = [\![B : A]\!]. \tag{B.18}$$

We recall two further results [**BDH11**, Cor. 7.26, Cor. 7.27] that are crucial for the proof of Theorem 5.2. Let $\|.\|_2$ stand for the l^2-norm of a vector. Let $N \subset M \subset \mathbf{B}(H)$ be factors such that the inclusion $N \subset M$ has finite index. If $M \subset A \subset \mathbf{B}(H)$ is such that one of the two relative commutants $N' \cap A$ or $M' \cap A$ is a factor and the other has finite-dimensional center, then

$$\big\|[\![N' \cap A : M' \cap A]\!]\big\|_2 \;\le\; [\![M : N]\!]. \tag{B.19}$$

Similarly, if $N \subset M \subset \mathbf{B}(H)$ are factors with $N \subset M$ of finite index, and $A \subset M' \subset \mathbf{B}(H)$ is such that one of the two algebras $N \vee A$ or $M \vee A$ is a factor and the other has finite-dimensional center, then

$$\big\|[\![M \vee A : N \vee A]\!]\big\|_2 \;\le\; [\![M : N]\!]. \tag{B.20}$$

B.VIII. Functors between module categories

The first lemma below is well known ([**JS97**, Rem. 2.1.3. (iii)]). It is the main distinguishing feature of the representation theory of von Neumann algebras. Here, ℓ^2 stands for $\ell^2(\mathbb{N})$ (or possibly $\ell^2(X)$ for a set X of sufficiently large cardinality if the Hilbert spaces we deal with are not separable).

LEMMA B.21. *Let A be a von Neumann algebra and let H and K be faithful left A-modules. Then $H \otimes \ell^2 \cong K \otimes \ell^2$. In particular, any A-module is isomorphic to a direct summand of $H \otimes \ell^2$.*

Let A and B be von Neumann algebras. We call a functor $F : A\text{-modules} \to B\text{-modules}$ *normal* if it is continuous with respect to the ultra-weak topology on hom-spaces, preserves adjoints $F(f^*) = F(f)^*$, and is additive in the following sense: for A-modules M_i the map $\oplus F(\iota_i)\colon \oplus F(M_i) \to F(\oplus M_i)$ induced by the inclusions $\iota_i\colon M_i \to \oplus_k M_k$ is a unitary isomorphism. Such functors are uniquely determined by their value on a single faithful A-module:

LEMMA B.22. *Let A and B be von Neumann algebras. Let M be a faithful A-module, let N be an arbitrary B-module, and let*

$$F_1 : \mathrm{End}_A(M) \to \mathrm{End}_B(N)$$

be a morphism of von Neumann algebras. Then the assignment $F(M) := N$, $F(f) := F_1(f)$ extends uniquely (up to unique unitary isomorphism) to a normal functor F from the category of A-modules to the category of B-modules.

PROOF. We prove existence and leave uniqueness to the reader. Given an A-module H, by Lemma B.21 we may pick an isomorphism

$$H \cong \operatorname{im}\big(p : M \otimes \ell^2 \to M \otimes \ell^2\big) \tag{B.23}$$

of H with the image of a projection $p \in \operatorname{End}_A(M) \mathbin{\bar{\otimes}} \mathbf{B}(\ell^2)$. We can then define

$$F(H) := \operatorname{im}\big((F_1 \otimes \operatorname{Id}_{\ell^2})(p) : N \otimes \ell^2 \to N \otimes \ell^2\big),$$

For morphisms, if $H \cong \operatorname{im}(p)$ and $K \cong \operatorname{im}(q)$ are A-modules given as above, then the image under F of an A-linear map $r : H \to K$ is the unique map $F(r) : F(H) \to F(K)$ for which the composite

$$N \otimes \ell^2 \twoheadrightarrow F(H) \xrightarrow{F(r)} F(K) \hookrightarrow N \otimes \ell^2$$

is the image under $F_1 \otimes \operatorname{Id}_{\ell^2}$ of the map $M \otimes \ell^2 \twoheadrightarrow H \xrightarrow{r} K \hookrightarrow M \otimes \ell^2$. □

A similar result holds for natural transformations.

LEMMA B.24. *Let F, G : A-modules $\to$ B-modules be two normal functors and let M be a faithful A-module. Then, in order to uniquely define a natural transformation $a : F \to G$, it is enough to specify its value on M, and to check that for each $r \in \operatorname{End}_A(M)$, the diagram*

$$\begin{array}{ccc} F(M) & \xrightarrow{F(r)} & F(M) \\ {\scriptstyle a_M}\downarrow & & \downarrow{\scriptstyle a_M} \\ G(M) & \xrightarrow{G(r)} & G(M) \end{array}$$

commutes.

PROOF. Given an A-module H along with an isomorphism (B.23), one uses the natural inclusion $F(H) \subset F(M \otimes \ell^2) \cong F(M) \otimes \ell^2$ to define

$$a_H := (a_M \otimes \operatorname{Id}_{\ell^2})|_{F(H)}.$$

This prescription is independent of the choice of isomorphism. □

B.IX. The split property

DEFINITION B.25. Given two commuting von Neumann algebras A and B acting on a Hilbert space H, we say that A and B are *split on* H if the natural map $A \otimes_{alg} B \to \mathbf{B}(H)$ extends to a homomorphism $A \mathbin{\bar{\otimes}} B \to \mathbf{B}(H)$. We also say that an inclusion $A_0 \hookrightarrow A$ is split if there exists a (equivalently, for any) faithful A-module H such that A_0 and A' are split on H.

LEMMA B.26. *Let $A_0 \subseteq A$ be von Neumann algebras acting faithfully on a Hilbert space H, and let $B \subseteq A'$ be an algebra that commutes with A. If the inclusion $A_0 \hookrightarrow A$ is split, then we have*

$$B \vee (A \cap A_0') = (B \vee A) \cap A_0'. \tag{B.27}$$

PROOF. Consider $H \otimes L^2(A_0)$ as an $A' \mathbin{\bar{\otimes}} A_0$-module, where A' acts on the first factor and A_0 acts on the second factor. Both H and $H \otimes L^2(A_0)$ are faithful $A' \mathbin{\bar{\otimes}} A_0$-modules.

So we may pick an $A' \bar{\otimes} A_0$-module isomorphism between $H \otimes \ell^2$ and $H \otimes L^2(A_0) \otimes \ell^2$. Let $K := L^2(A_0) \otimes \ell^2$, so we have $H \otimes \ell^2 \cong H \otimes K$. Under this identification, the subalgebra

$$\big(B \vee (A \cap A_0')\big) \bar{\otimes} \mathbf{B}(\ell^2) = (B \bar{\otimes} 1) \vee \big((A \bar{\otimes} \mathbf{B}(\ell^2)) \cap (A_0 \bar{\otimes} 1)'\big)$$

of $\mathbf{B}(H \otimes \ell^2)$ corresponds to

$$(B \bar{\otimes} 1) \vee \big((A \bar{\otimes} \mathbf{B}(K)) \cap (1 \bar{\otimes} A_0)'\big) = (B \bar{\otimes} 1) \vee (A \bar{\otimes} A_0') = (B \vee A) \bar{\otimes} A_0'$$

in $\mathbf{B}(H \otimes K)$. Similarly,

$$\big((B \vee A) \cap A_0'\big) \bar{\otimes} \mathbf{B}(\ell^2) = \big((B \bar{\otimes} 1) \vee (A \bar{\otimes} \mathbf{B}(\ell^2))\big) \cap (A_0 \bar{\otimes} 1)'$$

corresponds to

$$\big((B \bar{\otimes} 1) \vee (A \bar{\otimes} \mathbf{B}(K))\big) \cap (1 \bar{\otimes} A_0)' = \big((B \vee A) \bar{\otimes} \mathbf{B}(K)\big) \cap \big(\mathbf{B}(H) \bar{\otimes} A_0'\big) = (B \vee A) \bar{\otimes} A_0'.$$

The algebras (B.27) agree after tensoring with $\mathbf{B}(\ell^2)$, so they are equal. □

LEMMA B.28. *Let $A \subset \mathbf{B}(H)$ be a factor, and let A_0 be a subalgebra of A. If the inclusion $A_0 \hookrightarrow A$ is split, then*

$$(A_0 \vee A') \cap A = A_0. \tag{B.29}$$

PROOF. As in the previous lemma, we pick an isomorphism $H \otimes \ell^2 \cong H \otimes K$ of $A' \bar{\otimes} A_0$-modules, where $K = L^2(A_0) \otimes \ell^2$. Under that isomorphism, the algebras $A_0 \bar{\otimes} 1$ and

$$\big((A_0 \vee A') \cap A\big) \bar{\otimes} 1 = \big((A_0 \bar{\otimes} 1) \vee (A' \bar{\otimes} 1)\big) \cap (A \bar{\otimes} \mathbf{B}(\ell^2))$$

correspond to $1 \bar{\otimes} A_0$ and

$$\big((1 \bar{\otimes} A_0) \vee (A' \bar{\otimes} 1)\big) \cap (A \bar{\otimes} \mathbf{B}(K)) = (A' \bar{\otimes} A_0) \cap (A \bar{\otimes} \mathbf{B}(K)) = 1 \bar{\otimes} A_0.$$

Since their images in $\mathbf{B}(H \otimes K)$ agree, the two algebras (B.29) are equal. □

Recall the fiber product operation $*$ from Definition B.10.

LEMMA B.30. *Let A, B, and C be factors, and let $A^{op} \leftarrow C \rightarrow B$ be homomorphisms. Let $B \subset \hat{B}$ be a subfactor of finite index. Assuming that the inclusion $C \rightarrow A^{op}$ is split, then $A *_C \hat{B}$ and $A *_C B$ are factors, and we have*

$$[A *_C \hat{B} : A *_C B] = [\hat{B} : B]. \tag{B.31}$$

PROOF. Let H be a faithful A-module and K a faithful $\hat{B}$-module. Let A' be the commutant of A on H, and let B' and $\hat{B}'$ be the commutants of B and $\hat{B}$ on K. Finally, let C' be the commutant of C on K, and let $'C$ be the commutant of C^{op} on H.

Since the inclusion of C into A^{op} is split, so is the inclusion $A' \hookrightarrow {}'C$. The algebra C' is $'C$'s commutant on $H \boxtimes_C K$, and so A' and C' are split on $H \boxtimes_C K$. Finally, B' and $\hat{B}'$ being subalgebras of C', we conclude that A' and B', and also A' and $\hat{B}'$, are split on $H \boxtimes_C K$. It follows that the algebras

$$(A *_C B)' = A' \vee B' = A' \bar{\otimes} B' \quad \text{and} \quad (A *_C \hat{B})' = A' \vee \hat{B}' = A' \bar{\otimes} \hat{B}'$$

are factors, and thus so are $A *_C B$ and $A *_C \hat{B}$. Finally, we have

$$[A *_C \hat{B} : A *_C B] = [(A *_C B)' : (A *_C \hat{B})'] = [A' \bar{\otimes} B' : A' \bar{\otimes} \hat{B}'] = [B' : \hat{B}'] = [\hat{B} : B]$$

by (B.16) and (B.18). □

LEMMA B.32. *Let A_0 and A_1 be commuting subalgebras of $\mathbf{B}(H)$, and let B_0 and B_1 be commuting subalgebras of $\mathbf{B}(K)$. Let $C^{op} \to A_0$ and $C \to B_0$ be injective homomorphisms. If C^{op} and A_0' are split on H, then we have*

$$A_1 \vee (A_0 *_C B_0) \vee B_1 = (A_1 \vee A_0) *_C (B_0 \vee B_1) \tag{B.33}$$

on $H \boxtimes_C K$.

PROOF. As in the proof of the previous lemma, the algebras A_0' and C' are split on $H \boxtimes_C K$. In particular, the actions of A_0' on H and of B_0' on K induce an action of $A_0' \bar{\otimes} B_0'$ on $H \boxtimes_C K$.

Consider $H \otimes K$ as an $A_0' \bar{\otimes} B_0'$-module, where A_0' acts on H and B_0' acts on K. Since this is a faithful module, we can find an $A_0' \bar{\otimes} B_0'$-linear isometry

$$H \boxtimes_C K \hookrightarrow H \otimes K \otimes \ell^2.$$

Let $p \in \mathbf{B}(H \otimes K \otimes \ell^2)$ be its range projection. Under the induced isomorphism

$$\alpha : \mathbf{B}(H \boxtimes_C K) \xrightarrow{\cong} p\Big(\mathbf{B}(H) \bar{\otimes} \mathbf{B}(K) \bar{\otimes} \mathbf{B}(\ell^2)\Big)p,$$

we have

$$\alpha(A_0') = \big(A_0' \bar{\otimes} \mathbb{C} \bar{\otimes} \mathbb{C}\big)p, \qquad \alpha(B_0') = \big(\mathbb{C} \bar{\otimes} B_0' \bar{\otimes} \mathbb{C}\big)p,$$
$$\alpha(A_1) = \big(A_1 \bar{\otimes} \mathbb{C} \bar{\otimes} \mathbb{C}\big)p, \qquad \alpha(B_1) = \big(\mathbb{C} \bar{\otimes} B_1 \bar{\otimes} \mathbb{C}\big)p.$$

Recalling the definition $A_0 *_C B_0 := (A_0' \vee B_0')'$, we then see that

$$\alpha(A_0 *_C B_0) = p\big(A_0 \bar{\otimes} B_0 \bar{\otimes} \mathbf{B}(\ell^2)\big)p.$$

Similarly, $\alpha((A_1 \vee A_0) *_C (B_0 \vee B_1)) = p\big((A_1 \vee A_0) \bar{\otimes} (B_0 \vee B_1) \bar{\otimes} \mathbf{B}(\ell^2)\big)p$, and equation (B.33) follows since

$$(A_1 \bar{\otimes} \mathbb{C} \bar{\otimes} \mathbb{C}) \vee (A_0 \bar{\otimes} B_0 \bar{\otimes} \mathbf{B}(\ell^2)) \vee (\mathbb{C} \bar{\otimes} B_1 \bar{\otimes} \mathbb{C}) = (A_1 \vee A_0) \bar{\otimes} (B_0 \vee B_1) \bar{\otimes} \mathbf{B}(\ell^2). \quad \square$$

B.X. Two-sided fusion on L^2-spaces

Let M be a von Neumann algebra, and let M_0 and A be two commuting subalgebras such that $M_0 \vee A = M$. Let H_A be a faithful right A-module, and let B be its commutant, acting on H on the left. Then H is naturally a B–A-bimodule, and its conjugate $\overline{H}$ is an A–B-bimodule. Consider the Hilbert space

$$\widehat{H} := H \boxtimes_A L^2(M) \boxtimes_A \overline{H},$$

which is a completion of $\hom(L^2 A_A, H_A) \otimes_A L^2(M) \otimes_A \hom({}_A L^2 A, {}_A\overline{H})$.

Let us denote by J_A and J_M the modular conjugations on $L^2 A$ and $L^2 M$. There is an antilinear involution $\widehat{J} : \widehat{H} \to \widehat{H}$ given by

$$\widehat{J}\,(\varphi \otimes \xi \otimes \psi) = \bar{\psi} \otimes J_M(\xi) \otimes \bar{\varphi},$$

where $\xi \in L^2 M$ is a vector, and for $\varphi \in \hom(L^2 A_A, H_A)$ and $\psi \in \hom({}_A L^2 A, {}_A\overline{H})$, the maps

$$\bar{\varphi} \in \hom({}_A L^2 A, {}_A\bar{H}), \qquad \bar{\psi} \in \hom(L^2 A_A, H_A)$$

are given by $\bar{\varphi} = I \circ \varphi \circ J_A$ and $\bar{\psi} = I \circ \psi \circ J_A$, where I is the identity map between H and $\overline{H}$. There are natural left and right actions of B on $\widehat{H}$ coming from its actions on H and $\overline{H}$. Moreover, the left and right actions of M on $L^2(M)$ induce actions of M_0 on $\widehat{H}$. The left and right actions of M_0 and B are interchanged (up to a star) by $\widehat{J}$, and so the algebra $\widehat{M} := M_0 \vee B$ generated by them in their left action on $\widehat{H}$ is isomorphic to the algebra generated by them in their right action on

$\widehat{H}$. From this discussion, we see that $\widehat{H}$ is an $\widehat{M}$–$\widehat{M}$-bimodule with an involution $\widehat{J}$ that satisfies $\widehat{J}(a\xi b) = b^*\widehat{J}(\xi)a^*$.

PROPOSITION B.34. *In the above situation, there is a canonical positive cone $\widehat{P}$ in $\widehat{H} := H \boxtimes_A L^2M \boxtimes_A \overline{H}$ such that $(\widehat{H}, \widehat{J}, \widehat{P})$ is a standard form for $\widehat{M} = M_0 \vee B$.*

In the following proof, as in Section B.VIII, ℓ^2 stands for $\ell^2(\mathbb{N})$, or perhaps $\ell^2(X)$ for a set X of sufficiently large cardinality. If H admits a cyclic vector for A then we can replace ℓ^2 by $\mathbb{C}$ everywhere, and the proof simplifies.

PROOF. Pick an A-linear isometry $u\colon H_A \to \ell^2 \otimes L^2(A)_A$ (Lemma B.21) and let

$$\bar{u} := (1 \otimes J_A) \circ u \circ I : {}_A\overline{H} \to \overline{\ell^2} \otimes L^2(A) \cong {}_AL^2(A) \otimes \overline{\ell^2}.$$

The endomorphism algebra of $\ell^2 \otimes L^2(A)_A$ can be identified with $\mathbf{B}(\ell^2) \mathbin{\bar{\otimes}} A$. In particular, the range projection $p := uu^*$ is in $\mathbf{B}(\ell^2) \mathbin{\bar{\otimes}} A$.

Let us define $M_1 := \mathbf{B}(\ell^2) \mathbin{\bar{\otimes}} M$, with associated standard form $(L^2M_1, J_{M_1}, P_{M_1})$ and let $q := p\, J_{M_1} p\, J_{M_1} \in \mathbf{B}(L^2M_1)$ or, equivalently, $q(\xi) := p\,\xi\,p$. Composing $u \boxtimes \mathrm{id}_{L^2(M)} \boxtimes \bar{u}$ with the obvious identifications $(\ell^2 \otimes L^2A) \boxtimes_A L^2M \boxtimes_A (L^2A \otimes \overline{\ell^2}) \cong \ell^2 \otimes L^2M \otimes \overline{\ell^2} \cong L^2M_1$, we get an isometry

$$\begin{aligned} v \,:\, \widehat{H} &= H \boxtimes_A L^2M \boxtimes_A \overline{H} \to L^2M_1 \\ v(\varphi \otimes \xi \otimes \psi) &= (u\,\varphi) \cdot \xi \cdot (\bar{u}\,\psi) \end{aligned}$$

with range projection $vv^* = q$. The resulting isomorphism $\widehat{H} \cong q(L^2M_1)$ intertwines $\widehat{J}$ and qJ_{M_1}, as can be seen from the commutativity of the following diagram:

$$\begin{array}{ccc} \varphi \otimes \xi \otimes \psi & \xrightarrow{\ \widehat{J}\ } & \bar{\psi} \otimes J_M(\xi) \otimes \bar{\varphi} \\ & & \downarrow v \\ \downarrow v & & (u\,\bar{\psi}) \cdot J_M(\xi) \cdot (\bar{u}\,\bar{\varphi}) \\ & & \| \\ (u\,\varphi) \cdot \xi \cdot (\bar{u}\,\psi) & \xrightarrow{\ J_{M_1}\ } & (\bar{u}\,\psi)^* \cdot J_M(\xi) \cdot (u\,\varphi)^* \end{array}$$

Here, the last equality holds because the preimage of $u\,\bar{\psi}$ under the left action map $\ell^2 \otimes A \to \hom(L^2A_A, \ell^2 \otimes L^2A_A)$ agrees with the preimage of $(\bar{u}\,\psi)^*$ under the right action map $\ell^2 \otimes A \to \hom({}_AL^2A \otimes \overline{\ell^2}, {}_AL^2A)$, and the preimage of $\bar{u}\,\bar{\varphi}$ under the right action map agrees with the preimage of $(u\,\varphi)^*$ under the left action map.

Recall that B is the commutant of A on H. In its action on $L^2M_1 = \ell^2 \otimes L^2M \otimes \overline{\ell^2}$, we have $B \equiv vBv^* = q(\mathbf{B}(\ell^2) \mathbin{\bar{\otimes}} A)q$, and so it follows that

$$\begin{aligned} \widehat{M} \equiv v\,\widehat{M}\,v^* &= v(M_0 \vee B)v^* \\ &= qM_0 \vee q(\mathbf{B}(\ell^2) \mathbin{\bar{\otimes}} A)q \qquad \text{(B.35)} \\ &= q\big(\mathbf{B}(\ell^2) \mathbin{\bar{\otimes}} (M_0 \vee A)\big)q = q\big(\mathbf{B}(\ell^2) \mathbin{\bar{\otimes}} M\big)q = q\,M_1\,q. \end{aligned}$$

Now by [**Haa75**, Lem. 2.6], we know that $\big(q(L^2M_1), qJ_{M_1}, q(P_{M_1})\big)$ is a standard form for qM_1q. Therefore, by letting $\widehat{P} := v^{-1}(q(P_{M_1}))$, we have that $(\widehat{H}, \widehat{J}, \widehat{P})$ is a standard form for $\widehat{M}$. □

LEMMA B.36. *Let M, M_0, A, $\widehat{M}$ be as in the previous lemma. Then if M is a factor, so is $\widehat{M}$.*

PROOF. We have seen in (B.35) that $\widehat{M} = q\,M_1\,q = q(\mathbf{B}(\ell^2)\,\bar{\otimes}\,M)q$. The result follows since corners of factors are factors. □

The isomorphism constructed by Proposition B.34 satisfies the following version of associativity. Let $M = M_0 \vee A_1 \vee A_2$ be a von Neumann algebra, where M_0, A_1, and A_2 are commuting subalgebras of M. Let H_i be faithful right A_i-modules, and let B_i be their commutants. Then we can form the Hilbert spaces

$$\widehat{H}_1 := H_1 \boxtimes_{A_1} L^2M \boxtimes_{A_1} \overline{H}_1 \quad \text{and} \quad \widehat{H}_2 := H_2 \boxtimes_{A_2} L^2M \boxtimes_{A_2} \overline{H}_2$$

on which the algebras $\widehat{M}_1 := M_0 \vee B_1 \vee A_2$ and $\widehat{M}_2 := M_0 \vee A_1 \vee B_2$ act. By Proposition B.34, we have canonical isomorphisms $\widehat{H}_1 \cong L^2\widehat{M}_1$ and $\widehat{H}_2 \cong L^2\widehat{M}_2$.

Furthermore, we can form the Hilbert spaces

$$\widehat{\widehat{H}}_1 := H_2 \boxtimes_{A_2} L^2\widehat{M}_1 \boxtimes_{A_2} \overline{H}_2 \quad \text{and} \quad \widehat{\widehat{H}}_2 := H_1 \boxtimes_{A_1} L^2\widehat{M}_2 \boxtimes_{A_1} \overline{H}_1,$$

on which the algebra $\widehat{\widehat{M}} := M_0 \vee B_1 \vee B_2$ acts. Again by Proposition B.34, we then have canonical isomorphisms $\widehat{\widehat{H}}_1 \cong L^2\widehat{\widehat{M}} \cong \widehat{\widehat{H}}_2$.

PROPOSITION B.37. *In the above situation, the following diagram is commutative:*

$$\begin{array}{ccc} H_1 \underset{A_1}{\boxtimes} \widehat{H}_2 \underset{A_1}{\boxtimes} \overline{H}_1 & \longrightarrow & H_1 \underset{A_1}{\boxtimes} L^2\widehat{M}_2 \underset{A_1}{\boxtimes} \overline{H}_1 = \widehat{\widehat{H}}_2 \\ & & \downarrow \\ \cong \Big| & & L^2\widehat{\widehat{M}} \\ & & \uparrow \\ H_2 \underset{A_2}{\boxtimes} \widehat{H}_1 \underset{A_2}{\boxtimes} \overline{H}_2 & \longrightarrow & H_2 \underset{A_2}{\boxtimes} L^2\widehat{M}_1 \underset{A_2}{\boxtimes} \overline{H}_2 = \widehat{\widehat{H}}_1 \end{array} \tag{B.38}$$

PROOF. Let ℓ_1 and ℓ_2 be two copies of ℓ^2. Pick isometries $u_i\colon (H_i)_{A_i} \hookrightarrow (\ell_i \otimes L^2A_i)_{A_i}$, so as to identify $\widehat{H}_1$ with

$$L^2(p_1(\mathbf{B}(\ell_1)\,\bar{\otimes}\,M)p_1),$$

and $\widehat{H}_2$ with $L^2(p_2(M\,\bar{\otimes}\,\mathbf{B}(\ell_2))p_2)$, for $p_i := u_iu_i^*$. Here, we have $p_1 \in \mathbf{B}(\ell_1)\,\bar{\otimes}\,M$ and $p_2 \in M\,\bar{\otimes}\,\mathbf{B}(\ell_2)$. Let us also define the projections q_1 on $L^2(\mathbf{B}(\ell_1)\,\bar{\otimes}\,M) \cong \ell_1 \otimes \overline{\ell_1} \otimes L^2M$ and q_2 on $L^2(M\,\bar{\otimes}\,\mathbf{B}(\ell_2)) \cong L^2M \otimes \ell_2 \otimes \overline{\ell_2}$ by $q_i(\xi) = p_i\xi p_i$.

Given the above notations, consider the following diagram:

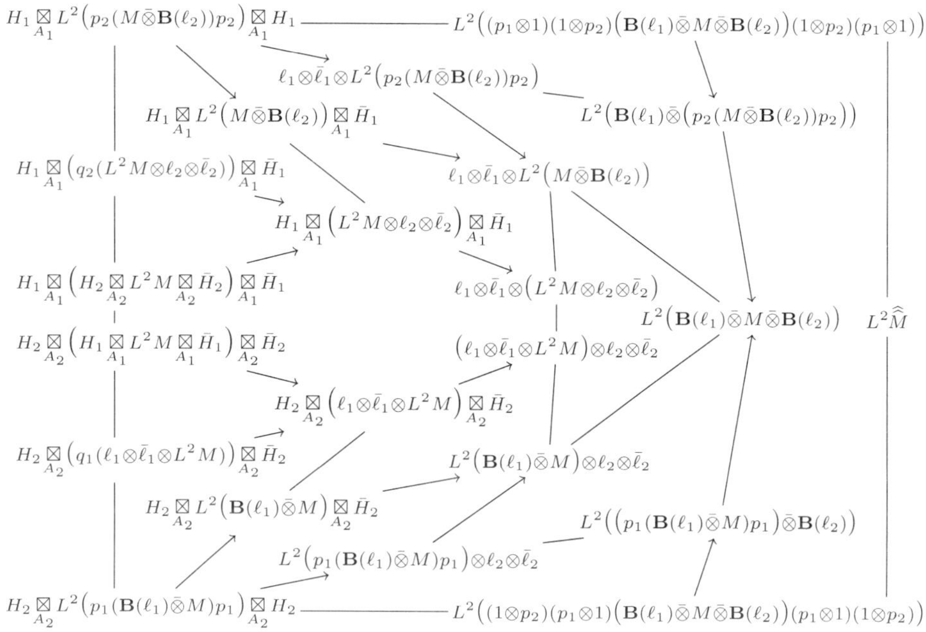

Here, arrows denote inclusions and lines denote isomorphisms. One recognizes (B.38) as the outside of the above diagram, and each one of the interior cells commutes for obvious reasons. □

APPENDIX C

Conformal nets

C.I. Axioms for conformal nets

Let VN be the category whose objects are von Neumann algebras with separable preduals, and whose morphisms are $\mathbb{C}$-linear homomorphisms and $\mathbb{C}$-linear antihomomorphisms. A *net* is a covariant functor $\mathcal{A}\colon \mathsf{INT} \to \mathsf{VN}$ taking orientation-preserving embeddings to injective homomorphisms and orientation-reversing embeddings to injective antihomomorphisms. We call a net *continuous* if for any intervals I and J, the map $\mathrm{Hom}_{\mathsf{INT}}(I, J) \to \mathrm{Hom}_{\mathsf{VN}}(\mathcal{A}(I), \mathcal{A}(J))$, $\varphi \mapsto \mathcal{A}(\varphi)$ is continuous for the $\mathcal{C}^\infty$ topology on $\mathrm{Hom}_{\mathsf{INT}}(I, J)$ and Haagerup's u-topology on $\mathrm{Hom}_{\mathsf{VN}}(\mathcal{A}(I), \mathcal{A}(J))$ [**BDH13**, Appendix]. Given a subinterval $I \subseteq K$, we will often not distinguish between $\mathcal{A}(I)$ and its image in $\mathcal{A}(K)$.

A *conformal net* is a continuous net $\mathcal{A}$ subject to the following conditions. Here, I and J are subintervals of an interval K:

(i) *Locality:* If $I \subset K$ and $J \subset K$ have disjoint interiors, then $\mathcal{A}(I)$ and $\mathcal{A}(J)$ are commuting subalgebras of $\mathcal{A}(K)$.

(ii) *Strong additivity:* If $K = I \cup J$, then $\mathcal{A}(K)$ is generated as a von Neumann algebra by the two subalgebras: $\mathcal{A}(K) = \mathcal{A}(I) \vee \mathcal{A}(J)$.

(iii) *Split property:* If $I \subset K$ and $J \subset K$ are disjoint, then the map from the algebraic tensor product $\mathcal{A}(I) \otimes_{alg} \mathcal{A}(J) \to \mathcal{A}(K)$ extends to a map from the spatial tensor product $\mathcal{A}(I) \bar{\otimes} \mathcal{A}(J) \to \mathcal{A}(K)$.

(iv) *Inner covariance:* If $\varphi \in \mathrm{Diff}_+(I)$ restricts to the identity in a neighborhood of ∂I, then $\mathcal{A}(\varphi)$ is an inner automorphism of $\mathcal{A}(I)$. (A unitary $u \in \mathcal{A}(I)$ with $\mathrm{Ad}(u) = \mathcal{A}(\varphi)$ is said to *implement* φ.)

(v) *Vacuum sector:* Suppose that $J \subsetneq I$ contains the boundary point $p \in \partial I$, and let $\bar{J}$ denote J with the reversed orientation; $\mathcal{A}(J)$ acts on $L^2(\mathcal{A}(I))$ via the left action of $\mathcal{A}(I)$, and $\mathcal{A}(\bar{J}) \cong \mathcal{A}(J)^{op}$ acts on $L^2(\mathcal{A}(I))$ via the right action of $\mathcal{A}(I)$. In this case, we require that the action of $\mathcal{A}(J) \otimes_{alg} \mathcal{A}(\bar{J})$ on $L^2(\mathcal{A}(I))$ extends to an action of $\mathcal{A}(J \cup_p \bar{J})$:

$$
\begin{array}{ccc}
\mathcal{A}(J) \otimes_{alg} \mathcal{A}(\bar{J}) & \longrightarrow & \mathbf{B}(L^2 \mathcal{A}(I)) \\
\downarrow & \nearrow & \\
\mathcal{A}(J \cup_p \bar{J}) & &
\end{array}
$$

Here, $J \cup_p \bar{J}$ is equipped with any smooth structure extending the given smooth structures on J and $\bar{J}$, and for which the orientation-reversing involution that exchanges J and $\bar{J}$ is smooth.

A conformal net $\mathcal{A}$ is called *irreducible* if the algebras $\mathcal{A}(I)$ are factors. As discussed in Correction 1.31, contrary to our claim in [**BDH13**, §1.4, Eq 1.42], we do not know whether an arbitrary conformal net decomposes as a direct integral of irreducible ones. A conformal net is called *semisimple* if it is a finite direct sum of irreducible conformal nets. We denote by CN_0 the symmetric monoidal category of semisimple conformal nets and their natural transformations. The tensor product of nets $\mathcal{A}$ and $\mathcal{B}$ is defined using the spatial tensor product of von Neumann algebras: $(\mathcal{A} \otimes \mathcal{B})(I) := \mathcal{A}(I) \bar{\otimes} \mathcal{B}(I)$. A natural transformation $\tau : \mathcal{A} \to \mathcal{B}$ between semisimple conformal nets is called *finite* if for all intervals I, the map $\tau_I : \mathcal{A}(I) \to \mathcal{B}(I)$ is a finite homomorphism (Appendix B.VI).

C.II. The vacuum sector

A *conformal circle* S is a circle S together with a diffeomorphism $S \to S^1$ that is specified up to a (not-necessarily orientation preserving) Möbius transformation of S^1 [**BDH13**, Def. 2.12]; here, S^1 denotes the standard circle $\{z \in \mathbb{C} : |z| = 1\}$. The set of conformal maps $S \to S'$ is denoted by $\mathrm{Conf}(S, S')$. If S and S' are oriented, then we denote by $\mathrm{Conf}_+(S, S')$ and $\mathrm{Conf}_-(S, S')$ the subsets of orientation preserving and orientation reversing maps. From now on, all our circles are implicitly assumed to be oriented.

For a conformal net $\mathcal{A}$ there is a functor [**BDH13**, Thm. 2.13]

$$S \mapsto H_0(S, \mathcal{A}) \tag{C.1}$$

from the category of oriented conformal circles to the category of Hilbert spaces. It sends orientation preserving conformal maps to unitary operators and orientation reversing conformal maps to anti-unitary operators. The Hilbert space $H_0(S, \mathcal{A})$ is called the *vacuum sector of* $\mathcal{A}$ *on* S, and comes equipped with compatible actions of the algebras $\mathcal{A}(I)$ for any subinterval I of S.

For $\varphi \in \mathrm{Conf}(S, S')$, the operator $H_0(\varphi, \mathcal{A})$ implements the diffeomorphism φ, that is:

$$\begin{aligned} \mathcal{A}(\varphi)(a) &= H_0(\varphi, \mathcal{A})\, a\, H_0(\varphi, \mathcal{A})^* \quad \text{if } \varphi \in \mathrm{Conf}_+(S, S') \\ \mathcal{A}(\varphi)(a) &= H_0(\varphi, \mathcal{A})\, a^*\, H_0(\varphi, \mathcal{A})^* \quad \text{if } \varphi \in \mathrm{Conf}_-(S, S') \end{aligned}$$

for any $I \subset S$ and $a \in \mathcal{A}(I)$.

Moreover, for every interval $I \subset S$, there is a canonical unitary identification

$$v_I \colon H_0(S, \mathcal{A}) \to L^2(\mathcal{A}(I)). \tag{C.2}$$

These unitaries are such that for $\varphi \in \mathrm{Conf}_+(S, S')$ and $\psi \in \mathrm{Conf}_-(S, S')$, the diagrams

(C.3)

$$\begin{array}{ccc} H_0(S,\mathcal{A}) & \xrightarrow{\;v_I\;} & L^2(\mathcal{A}(I)) \\ {\scriptstyle H_0(\varphi,\mathcal{A})}\downarrow & & \downarrow{\scriptstyle L^2(\mathcal{A}(\varphi))} \\ H_0(S',\mathcal{A}) & \xrightarrow{\;v_{\varphi(I)}\;} & L^2(\mathcal{A}(\varphi(I))) \end{array} \qquad \begin{array}{ccc} H_0(S,\mathcal{A}) & \xrightarrow{\;v_I\;} & L^2(\mathcal{A}(I)) \\ {\scriptstyle H_0(\psi,\mathcal{A})}\downarrow & & \downarrow{\scriptstyle L^2(\mathcal{A}(\psi j))\circ J} \\ H_0(S',\mathcal{A}) & \xrightarrow{\;v_{\psi(I')}\;} & L^2(\mathcal{A}(\psi(I'))) \end{array}$$

commute, where J is the modular conjugation on $L^2(\mathcal{A}(I))$, $j \in \mathrm{Conf}_-(S)$ is the involution that fixes ∂I, and $I' = j(I)$ is the closure of $S \setminus I$. Taking $\psi := j$ in the second diagram, we recover the modular conjugation as $J = v_I H_0(j_I, \mathcal{A}) v_I^*$.

If S is a circle without a conformal structure, then it is still possible to define $H_0(S, \mathcal{A})$ as $L^2(\mathcal{A}(I))$ of some interval $I \subset S$, but this only defines $H_0(S, \mathcal{A})$ up to non-canonical unitary isomorphism [**BDH13**, Def. 1.17]. We will sometimes abbreviate $H_0(S, \mathcal{A})$ by $H_0(S)$.

PROPOSITION C.4. (Haag duality for conformal nets [**BDH13**, Prop. 1.18]) *Let $\mathcal{A}$ be a conformal net, and let S be a circle. Then for any $I \subset S$, the algebra $\mathcal{A}(I')$ is the commutant of $\mathcal{A}(I)$ on $H_0(S, \mathcal{A})$.*

If $J \subset K$ are intervals such that J^c, the closure of $K \setminus J$, is itself an interval, then the relative commutant of $\mathcal{A}(J)$ in $\mathcal{A}(K)$ is $\mathcal{A}(J^c)$.

C.III. Gluing vacuum sectors

Consider a theta-graph Θ, and let S_1, S_2, S_3 be its three circle subgraphs with orientations as drawn below:

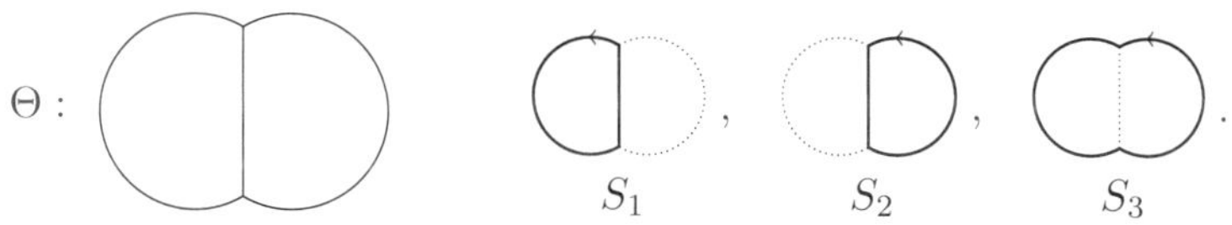

(Elsewhere in this book, we more often depict circles as squares:

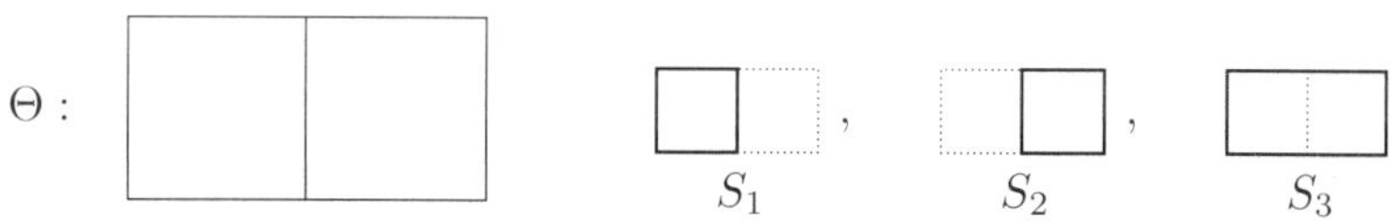

.)

We equip Θ with a 'smooth structure' in the sense of [**BDH14**, Def. 1.4] and let

$$I := S_1 \cap S_2, \quad K := S_1 \cap S_3, \quad L := S_2 \cap S_3.$$

Let us give K the orientation coming from S_1, and let us give I and L the orientations coming from S_2.

Then given a conformal net $\mathcal{A}$, there is a non-canonical isomorphism [**BDH13**, Cor. 1.34]

$$H_0(S_1, \mathcal{A}) \boxtimes_{\mathcal{A}(I)} H_0(S_2, \mathcal{A}) \;\cong\; H_0(S_3, \mathcal{A}), \tag{C.5}$$

compatible with the actions of $\mathcal{A}(K)$ and $\mathcal{A}(L)$. Moreover, in the presence of suitable conformal structures, this isomorphism can be constructed canonically: equip S_1 and S_2 with conformal structures, and let $j_1 \in \mathrm{Conf}_-(S_1)$, $j_2 \in \mathrm{Conf}_-(S_2)$ be the unique involutions fixing ∂I. Then there is a unique conformal structure on S_3 for which $j_2|_I \cup \mathrm{Id}_K : S_1 \to S_3$ and $j_1|_I \cup \mathrm{Id}_L : S_2 \to S_3$ are conformal. We can then use (C.2) to obtain the canonical isomorphism [**BDH13**, Cor. 2.20]

$$\begin{aligned} \Upsilon \colon H_0(S_1, \mathcal{A}) \boxtimes_{\mathcal{A}(I)} H_0(S_2, \mathcal{A}) &\xrightarrow{v_K \boxtimes v_I} L^2(\mathcal{A}(K)) \boxtimes_{\mathcal{A}(I)} L^2(\mathcal{A}(I)) \\ &\xrightarrow{\cong} L^2(\mathcal{A}(K)) \xrightarrow{v_K^*} H_0(S_3, \mathcal{A}). \end{aligned} \tag{C.6}$$

C.IV. Finite-index conformal nets

Let S be a circle, split into four intervals I_1, I_2, I_3, I_4 as follows:

(C.7)

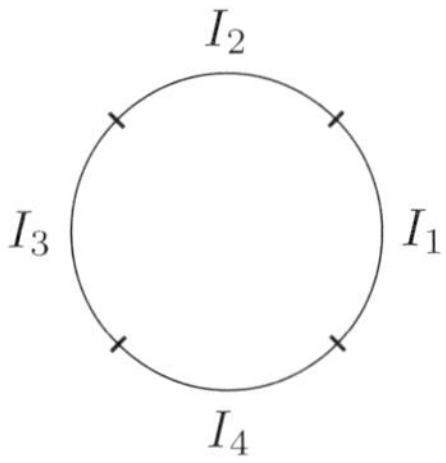

Given an irreducible conformal net $\mathcal{A}$, the algebras $\mathcal{A}(I_1 \cup I_3) = \mathcal{A}(I_1) \bar{\otimes} \mathcal{A}(I_3)$ and $\mathcal{A}(I_2 \cup I_4) = \mathcal{A}(I_2) \bar{\otimes} \mathcal{A}(I_4)$ act on $H_0(S, \mathcal{A})$ and commute with each other. The *index* $\mu(A)$ of $\mathcal{A}$ is then defined to be minimal index (see Appendix B.VII) of the inclusion $\mathcal{A}(I_1 \cup I_3) \subseteq \mathcal{A}(I_2 \cup I_4)'$ [**KLM01**, **Xu00**]:

$$\mu(\mathcal{A}) := [\mathcal{A}(I_2 \cup I_4)' : \mathcal{A}(I_1 \cup I_3)],$$

where the commutant is taken on $H_0(S, \mathcal{A})$.

C.V. Sectors and the Hilbert space of the annulus

Let $\mathcal{A}$ be an irreducible conformal net and let S be a circle (always oriented). An S-sector of $\mathcal{A}$ is a Hilbert space H together with homomorphisms

$$\rho_I \colon \mathcal{A}(I) \to \mathbf{B}(H), \quad I \subset S$$

subject to the compatibility condition $\rho_I|_{\mathcal{A}(J)} = \rho_J$ whenever $J \subset I$.

Let us write Δ for the set of isomorphism classes of irreducible S-sectors of $\mathcal{A}$. The vacuum sector discussed before is an example of a sector and we write 0 for the corresponding element of Δ. As all circles are diffeomorphic, Δ does not depend on the choice of circle S. There is an involution $\lambda \mapsto \bar{\lambda}$ on Δ given by sending an S-sector to its pull back along an orientation reversing diffeomorphism of S, as defined in [**BDH13**, (1.13)]. For $\lambda \in \Delta$, we write $H_\lambda(S)$ for a representative of λ as an S-sector. Of course, $H_\lambda(S)$ is only determined up to non-canonical isomorphism.

Let S_l be a circle, decomposed into four intervals $I_1, \ldots, I_4$ as in (C.7), and let S_r be another circle, similarly decomposed into four intervals $I_5, \ldots, I_8$. Let $\varphi \colon I_5 \to I_1$ and $\psi \colon I_7 \to I_3$ be orientation-reversing diffeomorphisms. These diffeomorphisms equip $H_0(S_l)$ with the structure of a right $\mathcal{A}(I_5) \bar{\otimes} \mathcal{A}(I_7)$-module. We are interested in the Hilbert space

$$H_\Sigma := H_0(S_l) \underset{\mathcal{A}(I_5)\bar{\otimes}\mathcal{A}(I_7)}{\boxtimes} H_0(S_r) \cong \underbrace{H_0(S_l) \boxtimes_{\mathcal{A}(I_5)} H_0(S_r) \boxtimes_{\mathcal{A}(I_3)}}_{\circlearrowleft}$$

This space is associated to the annulus $\Sigma = \mathbb{D}_l \cup_{I_5 \cup I_7} \mathbb{D}_r$, where $\mathbb{D}_l$ and $\mathbb{D}_r$ are disks bounding S_l and S_r. (As $H_0(S_l)$ and $H_0(S_r)$ are only determined up to non-canonical isometric isomorphism, the same Hilbert space H_Σ is, at this point, also only determined up to non-canonical isometric isomorphism.) Let $S_b := I_2 \cup I_8$ and $S_m := I_4 \cup I_6$ be the two boundary circles of this annulus.

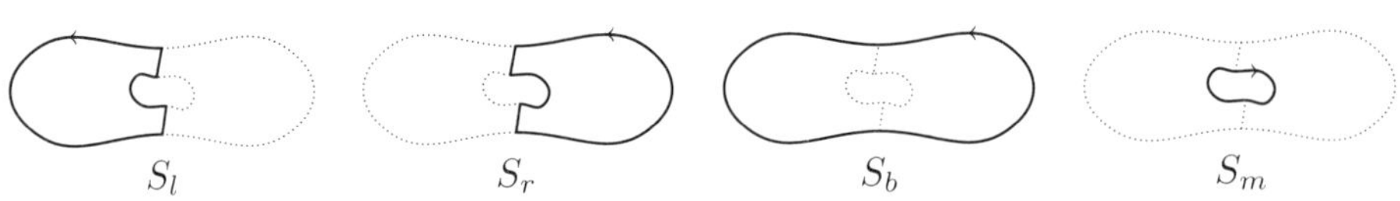

The Hilbert space H_Σ is an S_m-S_b-sector, which means that it is equipped with compatible actions of the algebras $\mathcal{A}(J)$ associated to all subintervals of S_m and S_b [**BDH13**, §3.2].

We finish by stating the computation of the annular Hilbert space, which, formulated in a different language, is due to [**KLM01**]:

THEOREM C.8 ([**BDH13**, Thm. 3.23, Thm. 3.14]). *If the conformal net $\mathcal{A}$ has finite index, then the set Δ is finite, and there is a unitary isomorphism of S_m-S_l-sectors*

$$H_\Sigma \cong \bigoplus_{\lambda\in\Delta} H_\lambda(S_m) \otimes H_{\bar\lambda}(S_b).$$

C.VI. Extension of conformal nets to all 1-manifolds

A priori, the only manifolds on which a conformal net $\mathcal{A}\colon \mathsf{INT} \to \mathsf{VN}$ can be evaluated are intervals. However, the functor $\mathcal{A}$ can be extended, in a canonical way, to the larger category $\mathsf{1MAN}$ of compact oriented 1-manifolds [**BDH14**, Thm. 1.3]. We denote the extension $\mathsf{1MAN} \to \mathsf{VN}$ by the same letter $\mathcal{A}$.

For S a circle, the algebra $\mathcal{A}(S)$ is defined to be the subalgebra of $\mathbf{B}(H_\Sigma)$ generated by $\mathcal{A}(I\times\{0\})$ for all $I \subset S$, where S is one of the two boundaries of the annulus $\Sigma := S\times[0,1]$.

THEOREM C.9 ([**BDH14**, Thm. 1.20]). *Let $\mathcal{A}$ be a conformal net with finite index and let S be a circle. Then there is a canonical isomorphism*

$$\mathcal{A}(S) \cong \bigoplus_{\lambda\in\Delta} \mathbf{B}(H_\lambda(S)). \tag{C.10}$$

Note that even though $H_\lambda(S)$ is only defined up to non-canonical isomorphism, its algebra of bounded operators is defined up to canonical isomorphism. It therefore makes sense for the isomorphism (C.10) to be canonical.

APPENDIX D

Diagram of dependencies

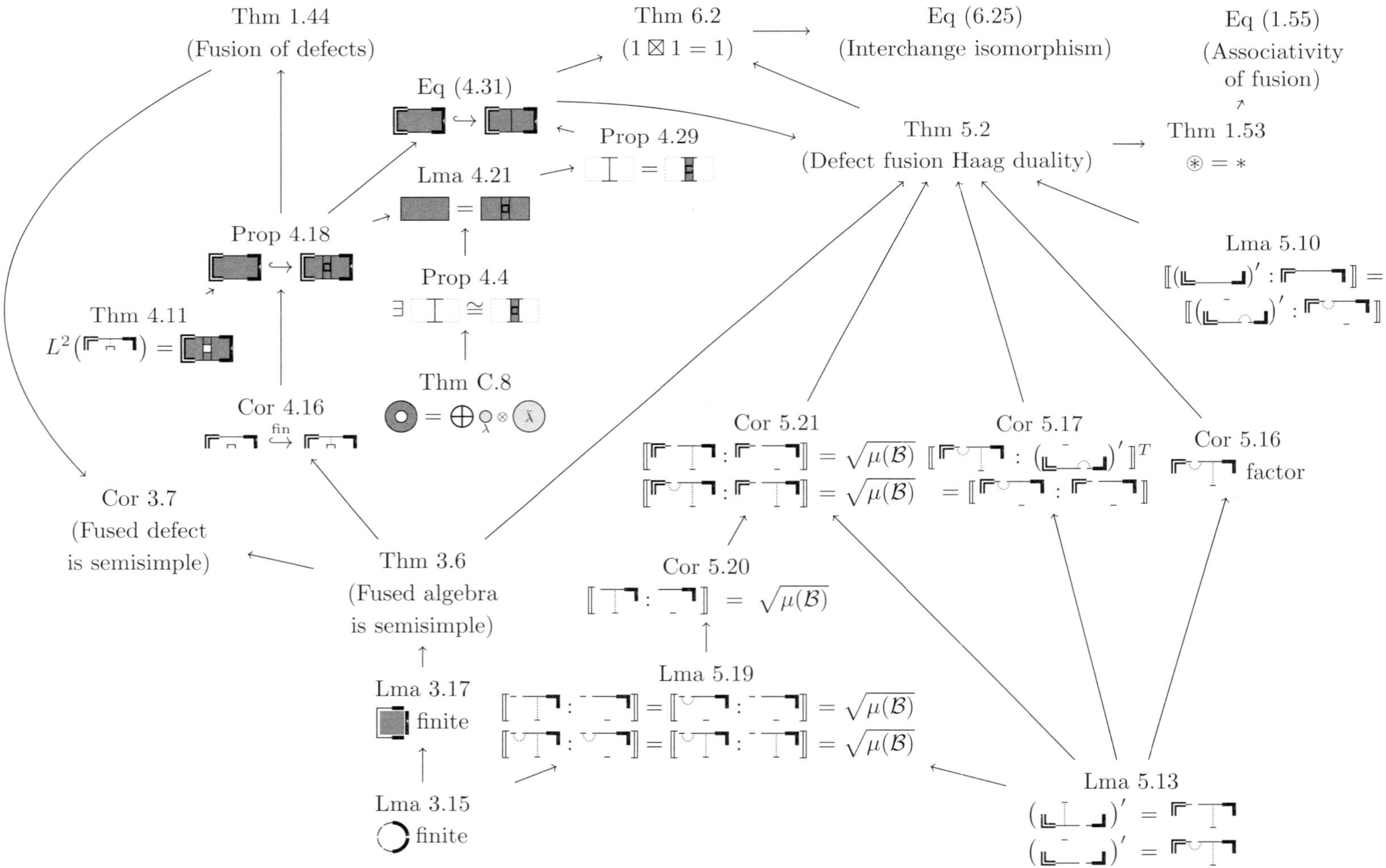

Thm 1.44
(Fusion of defects)
Thm 6.2
(1 ⊠ 1 = 1)
Eq (6.25)
(Interchange isomorphism)
Eq (1.55)
(Associativity
of fusion)
Eq (4.31)
Prop 4.29
Thm 5.2
(Defect fusion Haag duality)
Thm 1.53
⊛ = ∗
Lma 4.21
Prop 4.18
Prop 4.4
Thm 4.11
Lma 5.10
Cor 4.16
fin
Thm C.8
Cor 5.21
Cor 5.17
Cor 5.16
factor
Cor 3.7
(Fused defect
is semisimple)
Thm 3.6
(Fused algebra
is semisimple)
Cor 5.20
Lma 5.19
Lma 3.17
finite
Lma 3.15
finite
Lma 5.13

Bibliography

[BDH11] A. Bartels, C. L. Douglas, and A. Henriques, *Dualizability and index of subfactors*, Quantum Topol. **5** (2014), no. 3, 289–345. MR3342166

[BDH13] A. Bartels, C. L. Douglas, and A. Henriques, *Conformal nets I: Coordinate-free nets*, Int. Math. Res. Not. **13** (2015), 4975–5052, arXiv:1302.2604v2.

[BDH14] A. Bartels, C. L. Douglas, and A. Henriques, *Conformal nets II: Conformal blocks*, Comm. Math. Phys. **354** (2017), no. 1, 393–458, DOI 10.1007/s00220-016-2814-5. MR3656522

[BDH16] A. Bartels, C. L. Douglas, and A. Henriques, *Conformal nets IV: The 3-category*, arxiv:1605.00662, 2016.

[BE98] J. Böckenhauer and D. E. Evans, *Modular invariants, graphs and α-induction for nets of subfactors. I*, Comm. Math. Phys. **197** (1998), no. 2, 361–386. MR1652746

[BKLR14] M. Bischoff, Y. Kawahigashi, R. Longo, and K.-H. Rehren, *Phase boundaries in algebraic conformal QFT*, Comm. Math. Phys. **342** (2016), no. 1, 1–45. MR3455144

[BMT88] D. Buchholz, G. Mack, and I. Todorov, *The current algebra on the circle as a germ of local field theories*, Nuclear Phys. B Proc. Suppl. **5B** (1988), 20–56. Conformal field theories and related topics (Annecy-le-Vieux, 1988). MR1002955

[Con90] A. Connes, *Géométrie non commutative* (French), InterEditions, Paris, 1990. MR1079062

[DH12] C. L. Douglas and A. Henriques, *Internal bicategories*, arXiv:1206:4284, 2012.

[FSV13] J. Fuchs, C. Schweigert, and A. Valentino, *Bicategories for boundary conditions and for surface defects in 3-d TFT*, Comm. Math. Phys. **321** (2013), no. 2, 543–575. MR3063919

[GF93] F. Gabbiani and J. Fröhlich, *Operator algebras and conformal field theory*, Comm. Math. Phys. **155** (1993), no. 3, 569–640. MR1231644

[Haa75] U. Haagerup, *The standard form of von Neumann algebras*, Math. Scand. **37** (1975), no. 2, 271–283. MR0407615

[JS97] V. Jones and V. S. Sunder, *Introduction to subfactors*, London Mathematical Society Lecture Note Series, vol. 234, Cambridge University Press, Cambridge, 1997. MR1473221

[Kaw02] Y. Kawahigashi, *Generalized Longo-Rehren subfactors and α-induction*, Comm. Math. Phys. **226** (2002), no. 2, 269–287. MR1892455

[KL04] Y. Kawahigashi and R. Longo, *Classification of local conformal nets. Case $c < 1$*, Ann. of Math. (2) **160** (2004), no. 2, 493–522. MR2123931

[KLM01] Y. Kawahigashi, R. Longo, and M. Müger, *Multi-interval subfactors and modularity of representations in conformal field theory*, Comm. Math. Phys. **219** (2001), no. 3, 631–669. MR1838752

[Kos80] H. Kosaki, *Canonical L^p-spaces associated with an arbitrary abstract von Neumann algebra*, ProQuest LLC, Ann Arbor, MI, 1980. Thesis (Ph.D.)–University of California, Los Angeles. MR2630616

[Kos86] H. Kosaki, *Extension of Jones' theory on index to arbitrary factors*, J. Funct. Anal. **66** (1986), no. 1, 123–140. MR829381

[KS11] A. Kapustin and N. Saulina, *Surface operators in 3d topological field theory and 2d rational conformal field theory*, Mathematical foundations of quantum field theory and perturbative string theory, Proc. Sympos. Pure Math., vol. 83, Amer. Math. Soc., Providence, RI, 2011, pp. 175–198. MR2742429

[Lon94] R. Longo, *A duality for Hopf algebras and for subfactors. I*, Comm. Math. Phys. **159** (1994), no. 1, 133–150. MR1257245

[LR95] R. Longo and K.-H. Rehren, *Nets of subfactors*, Rev. Math. Phys. **7** (1995), no. 4, 567–597. Workshop on Algebraic Quantum Field Theory and Jones Theory (Berlin, 1994). MR1332979

[LX04] R. Longo and F. Xu, *Topological sectors and a dichotomy in conformal field theory*, Comm. Math. Phys. **251** (2004), no. 2, 321–364. MR2100058

[MT84] Y. Miura and J. Tomiyama, *On a characterization of the tensor product of self-dual cones associated to the standard von Neumann algebras*, Sci. Rep. Niigata Univ. Ser. A **20** (1984), 1–11. MR743064

[QRW07] T. Quella, I. Runkel, and G. M. T. Watts, *Reflection and transmission for conformal defects*, J. High Energy Phys. **4** (2007), 095, 45. MR2318788

[Sau83] J.-L. Sauvageot, *Sur le produit tensoriel relatif d'espaces de Hilbert* (French), J. Operator Theory **9** (1983), no. 2, 237–252. MR703809

[Sau85] J.-L. Sauvageot, *Produits tensoriels de $\mathcal{Z}$-modules et applications* (French), Operator algebras and their connections with topology and ergodic theory (Buşteni, 1983), Lecture Notes in Math., vol. 1132, Springer, Berlin, 1985, pp. 468–485. MR799587

[Seg04] G. Segal, *The definition of conformal field theory*, Topology, geometry and quantum field theory, London Math. Soc. Lecture Note Ser., vol. 308, Cambridge Univ. Press, Cambridge, 2004, pp. 421–577. MR2079383

[SFR06] C. Schweigert, J. Fuchs, and I. Runkel, *Categorification and correlation functions in conformal field theory*, International Congress of Mathematicians. Vol. III, Eur. Math. Soc., Zürich, 2006, pp. 443–458. MR2275690

[SW82] L. M. Schmitt and G. Wittstock, *Kernel representation of completely positive Hilbert-Schmidt operators on standard forms*, Arch. Math. (Basel) **38** (1982), no. 5, 453–458. MR666920

[Tak70] M. Takesaki, *Tomita's theory of modular Hilbert algebras and its applications*, Lecture Notes in Mathematics, Vol. 128, Springer-Verlag, Berlin-New York, 1970. MR0270168

[Tim08] T. Timmermann, *An invitation to quantum groups and duality*, EMS Textbooks in Mathematics, European Mathematical Society (EMS), Zürich, 2008. From Hopf algebras to multiplicative unitaries and beyond. MR2397671

[Was95] A. J. Wassermann, *Operator algebras and conformal field theory*, Proceedings of the International Congress of Mathematicians, Vol. 1, 2 (Zürich, 1994), Birkhäuser, Basel, 1995, pp. 966–979. MR1403996

[Was98] A. Wassermann, *Operator algebras and conformal field theory. III. Fusion of positive energy representations of* $\mathrm{LSU}(N)$ *using bounded operators*, Invent. Math. **133** (1998), no. 3, 467–538. MR1645078

[Xu00] F. Xu, *Jones-Wassermann subfactors for disconnected intervals*, Commun. Contemp. Math. **2** (2000), no. 3, 307–347. MR1776984

Editorial Information

To be published in the *Memoirs*, a paper must be correct, new, nontrivial, and significant. Further, it must be well written and of interest to a substantial number of mathematicians. Piecemeal results, such as an inconclusive step toward an unproved major theorem or a minor variation on a known result, are in general not acceptable for publication.

Papers appearing in *Memoirs* are generally at least 80 and not more than 200 published pages in length. Papers less than 80 or more than 200 published pages require the approval of the Managing Editor of the Transactions/Memoirs Editorial Board. Published pages are the same size as those generated in the style files provided for $\mathcal{A}_{\mathcal{M}}\mathcal{S}$-LaTeX or $\mathcal{A}_{\mathcal{M}}\mathcal{S}$-TeX.

Information on the backlog for this journal can be found on the AMS website starting from `http://www.ams.org/memo`.

A Consent to Publish is required before we can begin processing your paper. After a paper is accepted for publication, the Providence office will send a Consent to Publish and Copyright Agreement to all authors of the paper. By submitting a paper to the *Memoirs*, authors certify that the results have not been submitted to nor are they under consideration for publication by another journal, conference proceedings, or similar publication.

Information for Authors

Memoirs is an author-prepared publication. Once formatted for print and on-line publication, articles will be published as is with the addition of AMS-prepared frontmatter and backmatter. Articles are not copyedited; however, confirmation copy will be sent to the authors.

Initial submission. The AMS uses Centralized Manuscript Processing for initial submissions. Authors should submit a PDF file using the Initial Manuscript Submission form found at `www.ams.org/submission/memo`, or send one copy of the manuscript to the following address: Centralized Manuscript Processing, MEMOIRS OF THE AMS, 201 Charles Street, Providence, RI 02904-2294 USA. If a paper copy is being forwarded to the AMS, indicate that it is for *Memoirs* and include the name of the corresponding author, contact information such as email address or mailing address, and the name of an appropriate Editor to review the paper (see the list of Editors below).

The paper must contain a *descriptive title* and an *abstract* that summarizes the article in language suitable for workers in the general field (algebra, analysis, etc.). The *descriptive title* should be short, but informative; useless or vague phrases such as "some remarks about" or "concerning" should be avoided. The *abstract* should be at least one complete sentence, and at most 300 words. Included with the footnotes to the paper should be the 2010 *Mathematics Subject Classification* representing the primary and secondary subjects of the article. The classifications are accessible from `www.ams.org/msc/`. The Mathematics Subject Classification footnote may be followed by a list of *key words and phrases* describing the subject matter of the article and taken from it. Journal abbreviations used in bibliographies are listed in the latest *Mathematical Reviews* annual index. The series abbreviations are also accessible from `www.ams.org/msnhtml/serials.pdf`. To help in preparing and verifying references, the AMS offers MR Lookup, a Reference Tool for Linking, at `www.ams.org/mrlookup/`.

Electronically prepared manuscripts. The AMS encourages electronically prepared manuscripts, with a strong preference for $\mathcal{A}_{\mathcal{M}}\mathcal{S}$-LaTeX. To this end, the Society has prepared $\mathcal{A}_{\mathcal{M}}\mathcal{S}$-LaTeX author packages for each AMS publication. Author packages include instructions for preparing electronic manuscripts, samples, and a style file that generates the particular design specifications of that publication series. Though $\mathcal{A}_{\mathcal{M}}\mathcal{S}$-LaTeX is the highly preferred format of TeX, author packages are also available in $\mathcal{A}_{\mathcal{M}}\mathcal{S}$-TeX.

Authors may retrieve an author package for *Memoirs of the AMS* from `www.ams.org/journals/memo/memoauthorpac.html`. The *AMS Author Handbook* is available in PDF format from the author package link. The author package can also be obtained free of charge by sending email to tech-support@ams.org or from the Publication Division, American Mathematical Society, 201 Charles St., Providence, RI 02904-2294, USA. When requesting an author package, please specify $\mathcal{A}_{\mathcal{M}}\mathcal{S}$-LaTeX or $\mathcal{A}_{\mathcal{M}}\mathcal{S}$-TeX and the publication in which your paper will appear. Please be sure to include your complete mailing address.

After acceptance. The source files for the final version of the electronic manuscript should be sent to the Providence office immediately after the paper has been accepted for publication. The author should also submit a PDF of the final version of the paper to the editor, who will forward a copy to the Providence office.

Accepted electronically prepared files can be submitted via the web at www.ams.org/submit-book-journal/, sent via FTP, or sent on CD to the Electronic Prepress Department, American Mathematical Society, 201 Charles Street, Providence, RI 02904-2294 USA. TeX source files and graphic files can be transferred over the Internet by FTP to the Internet node ftp.ams.org (130.44.1.100). When sending a manuscript electronically via CD, please be sure to include a message indicating that the paper is for the *Memoirs*.

Electronic graphics. Comprehensive instructions on preparing graphics are available at www.ams.org/authors/journals.html. A few of the major requirements are given here.

Submit files for graphics as EPS (Encapsulated PostScript) files. This includes graphics originated via a graphics application as well as scanned photographs or other computer-generated images. If this is not possible, TIFF files are acceptable as long as they can be opened in Adobe Photoshop or Illustrator.

Authors using graphics packages for the creation of electronic art should also avoid the use of any lines thinner than 0.5 points in width. Many graphics packages allow the user to specify a "hairline" for a very thin line. Hairlines often look acceptable when proofed on a typical laser printer. However, when produced on a high-resolution laser imagesetter, hairlines become nearly invisible and will be lost entirely in the final printing process.

Screens should be set to values between 15% and 85%. Screens which fall outside of this range are too light or too dark to print correctly. Variations of screens within a graphic should be no less than 10%.

Any graphics created in color will be rendered in grayscale for the printed version unless color printing is authorized by the Managing Editor and the Publisher. In general, color graphics will appear in color in the online version.

Inquiries. Any inquiries concerning a paper that has been accepted for publication should be sent to memo-query@ams.org or directly to the Electronic Prepress Department, American Mathematical Society, 201 Charles St., Providence, RI 02904-2294 USA.

Editors

For a complete list of titles in this series, visit the
AMS Bookstore at **www.ams.org/bookstore/memoseries/**.